Enhancing Learning Through Technology

Elsebeth Korsgaard Sorensen,
Aalborg University, Denmark

Daithí Ó Murchú,
Gaelscoil Ó Doghair &
Innovative e-Learning/e-Tutoring, Hibernia College, Ireland

Information Science Publishing

Hershey • London • Melbourne • Singapore

Acquisitions Editor:	Michelle Potter
Development Editor:	Kristin Roth
Senior Managing Editor:	Amanda Appicello
Managing Editor:	Jennifer Neidig
Copy Editor:	Angela Thor
Typesetter:	Cindy Consonery
Cover Design:	Lisa Tosheff
Printed at:	Integrated Book Technology

Published in the United States of America by
Information Science Publishing (an imprint of Idea Group Inc.)
701 E. Chocolate Avenue
Hershey PA 17033
Tel: 717-533-8845
Fax: 717-533-8661
E-mail: cust@idea-group.com
Web site: http://www.idea-group.com

and in the United Kingdom by
Information Science Publishing (an imprint of Idea Group Inc.)
3 Henrietta Street
Covent Garden
London WC2E 8LU
Tel: 44 20 7240 0856
Fax: 44 20 7379 0609
Web site: http://www.eurospanonline.com

Library of Congress Cataloging-in-Publication Data

Enhancing learning through technology / Editors Elsebeth Korsgaard Sorensen and Daithi O Murchu.
    p. cm.
    Summary: "This book identifies and presents the latest research on theory, practice, and capturing learning designs and best-practices in education"--Provided by publisher.
    Includes bibliographical references and index.
    ISBN 1-59140-971-3 (hardcover) -- ISBN 1-59140-972-1 (softcover) -- ISBN 1-59140-973-X (ebook)
    1. Educational technology. 2. Learning. I. Sorensen, Elsebeth Korsgaard. II. Murchu, Daithi O
    LB1028.3.E64 2006
    371.33--dc22
                                        2006003555

British Cataloguing in Publication Data
A Cataloguing in Publication record for this book is available from the British Library.

Enhancing Learning Through Technology

Table of Contents

Preface

Digital technology and learning processes have become interrelated entities. Entities that, in principle, through the interconnectivity of their relationships, give birth to a new phenomenon: "Learning-through and with digital technology." As relatively new as it is, our expectations, over the years, to the qualitative potential of this phenomenon for education have been high (Collis, 1996; Harasim, Hiltz, Teles, & Turoff, 1995; Mason, 1998; Sorensen & Takle, 2002). In contrast, the fact that we still identify and name it by its separate titles suggests that our understanding of its nature is still evolving, and much research on practice and implementation in education indicates that digital technology and learning, within pedagogical design and delivery, in many cases, are perceived and treated as the two, separate, constituting entities. Regardless of our ability to grasp the interwoven nature of this new phenomenon, learning today, through, and with, digital technology, is the new global reality, and it is offering us an entirely new educational paradigm. As time progresses, researchers and practitioners, from each of their perspectives, are still struggling to grasp and understand the implications for education, and mobilise — in holistic and integrated ways–the latent potential of the new educational paradigm, in order to enhance and make processes of learning through technology genuine, joyful, meaningful, social, and engaging.

The new paradigm imposes global challenges to education, at both macro- and microlevels. At a macrolevel, the challenges to be faced are broad and include digital divides, illiteracy, political challenges, and intercultural diversity (Brown & Davis, 2004), just to mention a few. In a microperspective, in the context of formal education, the alternative nature of this paradigm requires both techno-logical skills, and innovative pedagogical-methodological innovations and "re-imaginations" (Gibson, 2005) on behalf of teachers and learners. But such must be born and tied from a perspective of what constitutes learning as a genuine, joyful, social, and engaging endeavour, and in a confrontation with the complex challenges of the new educational paradigm.

For more than 2 decades, the potential of this new paradigm for education has been researched, practiced, praised, and scorned. It has been offered to us, challenged us as educators and educational designers, and presented its invita-tion to us for educational innovation and change. The challenge has been widely accepted and taken up, and many experiments and experiences aiming at utiliz-ing the new paradigm for teaching and learning processes have been made, some more successfully than others (Bates, 1999; Collis, 1996; Collis, 2001; Miyake & Koschmann, 2002). The vast majority, however, have not necessar-ily achieved their goal to actually "enhance" learning and the quality of the learning process, viewed from the perspective of the learner. Utilizing the po-tential of the new paradigm for learning, and cultivating learning processes of "good" quality, remains a controversial issue.

In the latter part of the twentieth century, much debate emerged concerning the definition, vision, reality, and functionality of the term technology. To this present day, the *Oxford Encyclopaedic English Dictionary* (1991) defines technol-ogy as being "the study or use of the mechanical arts and applied sciences": a rather narrow and tunnelled definition of what society today perceives and en-visions "technology" to be. One would be safe to say that citizens worldwide encamped in mainly one of four areas: those who regarded technology as the single greatest "medium" and "tool" of future possibility for society as a whole; those who held that technology, no matter how one defined it, would be the "damnation" of both humanity and society; those who remained on the side-lines, at times oblivious to what was taking place around them; and, finally, those who really did not care one way or the other. With the dot.com boom came a belief that "technology" was the tool through which all citizens of the world, and global economies, would be enhanced to levels that would benefit both developed and developing societies. Life would be easier, computers would make our quality of life better, fortunes would be made and lost, and people, in general, would have more quality time at home and at play. The personal com-puter, and its associate peripherals, infiltrated millions of homes, schools, of-fices, and institutions, and with the birth of the "Internet" came a new dawn: a dawn of synchronous and asynchronous access, and sharing without barriers and distance. The future looked wonderful, and governments implemented plans to plough millions of pounds, dollars, and Euros into educational institutions in the belief that the hardware would change the work practices and processes at all levels within society. Now, as we enter the early days of the 21st century, technology, as we know it, is without boundaries. The possibilities are vast, and the definitions are so varied that it would be suicidal to attempt to confine the term "technology" to just any one practice or meaning.

Which theoretical horizons does research, so far, suggest as having proved generally promising in terms of providing fruitful inspiration and insights, with the aim of "enhancing learning through technology"? While the term "learning," as well as research in learning, have gone through their stages of evolution, the insights achieved in this respect, for example, within the research fields of both computer-supported collaborative learning (CSCL) (Dillenbourg, Baker, Blaye, & O'Malley, 1995; Koschman, 1996; Miyake & Koschmann, 2002; O'Malley, 1995; Pea, 1996) and open and distance learning (ODL) (Tait, 2003), should be also noted.

In particular, within online and networked learning, the general and widely accepted approaches — as also demonstrated by many of the chapters in this book — are based on principles of collaborative learning (Roberts, 2004; Sorensen, 2004). As emphasized by Roberts (2004), collaborative learning is by no means a new concept. However, within research on collaborative learning, peer interaction, and the perception on its implication for the learning process, differs, and the different views are connected, theoretically, to different learning theorists (Dillenbourg et al., 1995). But whether we adapt a socioconstructivist, a sociocultural, or a shared cognition approach to collaborative learning, it is beyond any doubt that the essential power of the concept for enhancing online and networked learning through technology, should be sought in the fact that its principles are rooted in dialogue and interaction, and thus, focused in the very fundamental condition of human existence, and "medium" of human growth and prosperity (Heidegger, 1986; Sorensen, 2004).

Tony Kaye, back in 1992, described one of the challenges afforded by the new paradigm as a consequence of this, as the potential for "learning together apart" (Kaye, 1992, p. 1). To us, the editors, there is no doubt that regardless of the distance between learners, the learning potential of the new, educational paradigm is related to the basic human principle of collaborative dialogue and knowledge building, and to the collaboration enabled between learners (Ó Murchu & Sorensen, 2004; Salmon, 2000; Scardamalia & Bereiter, 1996; Sorensen & Takle, 2004).

Which criteria would we use to identify a learning process as a genuine, joyful, social, and engaging endeavour? In order for learners to develop profiles of critical and democratically-oriented global citizens, learning processes need to unfold joyfully, collaboratively, and nonauthoritatively, in a shared trajectory among learners, allowing them to fully engage:

When we do things with purpose and conviction, we recharge our vital batteries, and the sudden flash of 'life failure' never occurs. But when we begin to live mechanically, performing our everyday tasks as a mere habit, this robotic activity fails to recharge our vital batteries; then, suddenly, our inner resources are inadequate to meet some sudden challenge. (Wilson, 1998, p. 202)

The processes we, the editors, envision for learners deny the necessity for boredom, and refuse to accept the common state of a resulting "fatigue" in learners. Learning processes should not only be relevant and "educating" with respect to the development of democratic values and skills, they should also, in holistic ways, be conducive to genuine (Colaizzi, 1978), meaningful (Jonassen, Peck, & Wilson, 1999), and soulful learning (Sorensen & Ó Murchú, 2005), while cultivating motivation, initiative, ownership, and joyful engagement (Wenger, 1998; Wenger, McDermott, & Snyder, 2002) as part of their methodological considerations — the when, where, what, and how:

For to miss the joy is to miss all. The mystic sense of hidden meaning (...) the glow of meaning and purpose hidden inside everyone. (Wilson, 1998, p. 203)

How can we envision the embodiment of this new paradigm at work in achieving such a type of learning in the name of "enhancement of learning through technology"? What might potentially become the next subtle insight for educators and designers of education to uncover — or discover — and explore?

In our call for chapters for the book, we challenged the academic world to leave any fixed definitions of "learning" and "technology" aside. We invited the authors, in the presentation of their research, to "go outside the box" and, with open minds and souls, boldly envision what no one else had reported in the existing body of literature on enhancement of learning through technology. Based on their own experiences and insights, we asked them to write for the future, and to dream the dreams that would challenge others to engage in debate, and battle to finding vital future pathways of enhancing learning through technology. Learning, viewed as a joyful, holistic, democratic, and transformative process that unfolds collaboratively and genuinely, anywhere and anytime, in a framework and context that soulfully allows the learner to "breathe" or "energize":

Stating the thing broadly, the human individual lives far within his limits; he possesses powers of various sorts which he habitually fails to use. He energizes below his maximum, and he behaves below his optimum. (Wilson, 1998, p. 205)

In Chapter I, "Online Communities and Professional Teacher Learning: Affordances and Challenges," Norbert Pachler and Caroline Daly, from England, establish the contemporary context for developing teachers' professional learning through the affordances of new technologies. Their errand here is to establish what claims can be made about the potential of online communities to provide a counter to the reductive models of professional development that have dominated teachers' learning in England and Wales in recent years. Most critically, new technologies have a contribution to make in debating the professional and ethical principles that govern the choices teachers make about taking action that affects learning: their pupils', as well as their own. Only from this perspective, can technologies affect the development of a practice that is based on enhanced, critical understandings of what they "can do" rather than what they are told to do.

Anders Olafsson and J. Ola Lindberg from Sweden, in Chapter II, "Enhancing Phronesis: Bridging Communities Through Technology," explore the possibilities to use technology in order to improve the contextual and value-based dimensions in online, distance-based teacher training in Sweden. It is argued that active participation, collaboration, and dialogue are vital in order to foster common moral and societal values among the teacher trainees, but that there is a need for rethinking how technology could be used in order to accommodate such processes. The chapter suggests that a development of a shared teacher identity is possible by expanding the scope of online community, and bridging teacher-training practices to teacher practices, thus including already practicing teachers, teacher trainers, and teacher trainees in a joint educational community with the crucial input of technology.

In Chapter III, "Enhancing the Design of a Successful Networked Course Collaboration: An Outsider Perspective," Rema Nilakanta (USA), Laura Zurita (Denmark), Olatz López Fernandez (Spain), Elsebeth Korsgaard Sorensen (Denmark), and Eugene Takle (USA) present a preliminary critique of an online, transatlantic collaboration designed for collaborative learning. The critique by external reviewers, using qualitative methods within the interpretivist paradigm, hints at critical factors necessary for successful online collaborative learning. The evaluation supports the view that in order to raise the quality of online dialogue and enhance deep learning, it is good practice to heed, as well as give voice to participants' needs by involving them directly in the design of the course. With the proliferation of e-learning in higher education, it is important that we pay close attention to the design of online technologies and pedagogies that claim to support learning that is necessary for a global world: learning that aims to develop future leaders who are successful across cultures, disciplines, and geography. This requires not only a focus on the design of online courses, but also exploring innovative ways of evaluating them.

In Chapter IV, Ian Gibson, USA, in his chapter "Enhanced Learning and Leading in a Technology-Rich, 21st Century Global Learning Environment," explores the evolution of thinking about learning, resulting from the increasingly ubiquitous presence of instructional technology and communications technology in learning environments. The chapter further describes the impact of technology on the potential transformation of four-walled classrooms into global, online, learning communities from a constructivist perspective, while looking at learner/teacher roles in the learning process. Expectations for education are changing. The knowledge base of education is changing. Conceptions of how individual learning occurs are changing. The tools available to "do" education are changing. The roles of teachers are changing. Understandings of what should be learned, who should be learning, how they should learn, where they should learn, and when they should learn, are changing. So, expecting school leaders to recreate their conceptions of what constitutes appropriate leader behaviour should also change!

Erik Champion, from Australia, in Chapter V, titled "Enhancing Learning Through 3-D Virtual Environments," delves into an area of global debate, and clearly states that educators cannot begrudge students their envy in looking at popular films and computer games as major contenders for their spare time. While teachers could attempt to fight the popularity of games, he suggests a more useful endeavor would be to attempt to understand both the temptation of games, and to explore whether we could learn from them, in order to engage students to learn, and to educate them at the same time.

Chapter VI, "Inquisitivism: The Evolution of a Constructivist Approach for Web-Based Instruction," by Dwayne Harapnuik from Canada, introduces "inquisitivism" as an approach for designing and delivering Web-based instruction that shares many of the same principles of minimalism and other constructivist approaches. Inquisitivism is unique in that its two primary or first principles are the removal of fear and the stimulation of an inquisitive nature. The approach evolved during the design and delivery of an online, full-credit university course. The results of a quasi-experimental design-based study revealed that online students in the inquisitivism-based course scored significantly higher on their final project scores, showed no significant difference in their satisfaction with their learning experiences from their face-to-face (F2F) counterparts, and had a reduction in fear or anxiety toward technology. Finally, the results revealed that there was no significant difference in final project scores across the personality types tested. The author hopes that inquisitivism will provide a foundation for creating effective constructivist-based, online learning environments.

Pirkko Raudaskoski, from Denmark, in Chapter VII, "Situated Learning and Interacting With/Through Technology: Enhancing Research and Design," discusses the growing interest within social and humanistic sciences towards understanding theoretical and analytical practice. Lave and Wenger's concept, "situated learning," describes the process of newcomers moving toward full participation in a community. Situated learning is equalled with social order: instead of understanding learning as a separate practice from everyday life, learning is seen as a more mundane phenomenon. Ethnomethodology and conversation analysis (CA) find that social order is created continuously by its members in their interactions. As ethnomethodology and CA base their findings on rigorous data analysis, they are extremely useful in analysing situated learning in everyday practices. The interdisciplinary interaction analysis (IA) is suggested as the best way to study the various aspects of situated learning in technology-intensive interactions. Learning is taking place all the time, in educational and other institutions and in everyday life, and Pirkko boldly states that technology played a decisive role at all stages of the learning process.

Jørgen Bang and Christian Dalsgaard, both from Denmark, envision in Chapter VIII "Rethinking E-Learning: Shifting the Focus to Learning Activities," perspectives on rethinking e-learning, shifting the focus to learning activities. They clearly state that technology alone does not deliver educational success. It only becomes valuable in education if learners and teachers can do something useful with it. Their main goal is to rethink e-learning by shifting the focus of attention from learning resources (learning objects) to learning activities, which also implies a refocusing of the pedagogical discussion of the learning process. Firstly, they identify why e-learning has not been able to deliver the educational results as expected 5 years ago. Secondly, they discuss the relationship between learning objectives, learning resources, and learning activities, in an attempt to develop a consistent, theoretical framework for learning as an active collaborative process that bears social and cultural relevance for the student. Finally, they boldly specify their concept of learning activities, and argue for the educational advantages of creating large learning resources that may be used for multiple learning activities.

Chapter IX, "Empirical Analyses of Computer-Supported Collaborative Learning and the Central Research Questions: Two Illustrative Case Studies," is written by Tony Carr from South Africa, Vic Lally from theUK, Maarten de Laat from the UK, and Glenda Cox from South Africa. Their chapter examines the theoretical and conceptual issues involved in gathering evidence to build a database for the design of virtual higher education (computer supported collaborative learning — CSCL and networked learning — NL). After briefly surveying the current state of CSCL/NL research and its lack of theoretical synthesis, the authors cleverly propose three high-level research questions as a way of focusing efforts on finding answers. In particular, the authors look at the way theory and praxis (theory-informed practice) might be more effectively and boldly engaged through "theory-praxis conversations," in order to make effective use of empirical data to build the evidence base that will be needed to design and build virtual higher education over the next 10 years.

Chapter X, titled "Identifying an Appropriate Pedagogical Networked Architecture for Online Learning Communities within Higher and Continuing Education," is written by the editors, Elsebeth Korsgaard Sorensen (Denmark) and Daithí Ó Murchú (Ireland). It addresses the problem of enhancing the quality of online learning processes through pedagogic design. Based on their earlier research findings from analysis of two comparable online master courses offered in two Masters' programmes, respectively from Denmark and Ireland (Ó Murchú & Sorensen, 2004), they present what they boldly assert to be a fruitful, student-centred, pedagogic model for design of networked learning. The design model is composed of what they have identified as unique characteristics of online learning architectures that, in principle, promote and allow for global intercultural processes of meaningful learning through collaborative knowledge building in online communities of practice.

Chapter XI is written by John Cuthell from the UK, and is titled "Ms. Chips and Her Battle Against the Cyborgs: Embedding ICT in Educational Practice." This chapter focuses on practicing teachers, and examines the institutional and individual factors that inhibit the implementation of information and communication technology (ICT) as a tool for teaching and learning. The affordances of ICT are identified, together with their contribution to attainment, creativity, and learning. John argues that many of the obstacles to meaningful uses of ICT are embedded in the assumptions inherent in many institutional frameworks that are predicated on an outmoded industrial model. This model drives many school timetables that process learners through the school machine. Individual change is easier to effect than institutional: the author boldly provides suggestions to liberate creative teachers from constraints of the system.

Chapter XII, titled "Making Sense of Technologically Enhanced Learning in Context: A Research Agenda," is written by Simon B. Heilesen and Sisse Siggaard Jensen from Denmark. It proposes that technologically enhanced learning should be understood and evaluated by means of a combination of analytical strategies. These strategies will allow us to analyze it, both as seen from the macroanalytical or "outside" perspective of a rich, social, cultural, and technological context, and from a microanalytical or "inside out" perspective of individual sense making in learning situations. As a framework, Simon and Sisse use sense-making methodology, and a model for causal-layered analysis limited to the "remediated classroom" of constructivist, netbased, university education. Problematizing some common assumptions about technologically enhanced learning, the authors boldly define 10 questions that may serve as the basis for a research agenda meant to help to understand why the many visions and ideals of the online or remediated classroom are not more widely realized and demonstrated in educational design and practice.

The final chapter, XIII, titled "Brain-Based Learning," is written by Kathleen Cercone from the USA. Addressing an area of much debate, it explores the dynamic field of neuroscience research, which explains how the brain learns.

Since the 1990s, there has been explosive growth in information about the neurophysiology of learning. A discussion of the neuroanatomy that is necessary to understand this research is presented first. Kathy further describes current brain research, with particular focus on its implications for teaching adult students in an online environment. In addition, two instructional design theories (Gardner's Multiple Intelligence and Kovalik's Integrated Thematic Instruction), which have a basis in neuroscience, are further examined. Recommendations founded on brain-based research, with a focus on adult education, follow, including specific activities such as crossed-lateral movement patterns, and detailed online activities that can be incorporated into an online learning environment or a distance learning class (and face-to-face classrooms) for adults. Comprehensive recommendations and guidelines for online learning design have been provided as suggestions for making maximum use of the brain-based principles discussed in this chapter.

All 13 chapters address "enhancing learning through technology" in a manner that challenges the reader to morph current thinking, and they boldly suggest new pathways and avenues towards ensuring that technology, no matter how we define it, and no matter how we present it, becomes an integral, meaningful, and soulful collaborator at all levels of education and society.

The editors believe that it is no longer acceptable to regard learning simply as a product of teaching, and to regard technology as the study or use of the mechanical arts and applied sciences. Neither is it acceptable for educational practitioners and researchers to continue to treat the two phenomena of "digital technology" and "meaningful learning" as two separate entities, and thereby, indirectly avoid the confrontation with the necessity of facing the complex integral design challenges imposed by the new educational paradigm. We hope and believe that this book will cause many to debate, discuss, and invent future pathways of enhancing the quality of learning through digital technology, as a means of cultivating and promoting prosperity of genuine, meaningful, and soulful learning.

References

Bates, A. W. (1999). *Managing technological change: Strategies for academic readers*. San Francisco: Jossey Bass.

Brown, A., & Davis, N. (2004). Introduction. In A. Brown & N. Davis (Eds.), *Digital technology communities and education* (pp.1-12). London: RoutledgeFarmer.

Colaizzi, P. F. (1978). Learning and existence. In R. Valle & M. King (Eds.), *Existential-phenomenological alternatives for psychology* (pp. 119-135). New York: Oxford University Press.

Collis, B. (1996). *Tele-learning in a digital world: The future of distance learning*. London: International Thomson Publications.

Dillenbourg, P., Baker, M., Blaye, A., & O'Malley, C. (1995). The evolution of research on collaborative learning. In P. Reimann & H. Spada (Eds.), *Learning in human and machines. Towards an interdisciplinary learning science* (pp. 189-211). London: Pergamon.

Gibson, I. W. (2005, June 17). *The future ain't what it used to be!!* Graduation speech, Aalborg University, Denmark.

Harasim, L., Hiltz, S. R., Teles, L., & Turoff, M. (1995). *Learning networks*. Cambridge, MA: MIT Press.

Heidegger, M. (1986). *Sein und Zeit*. Tübingen: Max Niemeyer Verlag..

Jonassen, D. H., Peck, K. L., & Wilson, B. G. (1999). *Learning with technology: A constructivist perspective* . Upper Saddle River, NJ: Merrill Publishing.

Kaye, A. R. (1992). Learning together apart. In A.R. Kaye (Ed.), *Collaborative learning through computer conferencing* (pp. 1-12). Heidelberg: Springer-Verlag. NATO ASI series.

Koschmann, T. D. (1996). Paradigm shifts and instructional technology: An introduction. In T. D. Koschmann (Ed.), *CSCL: Theory and practice of an emerging paradigm* (pp. 1-23). Mahwah, NJ: Lawrence Erlbaum Associates Publishers.

Mason, R. (1998). *Globalising education. Trends and applications*. London: Routledge.

Miyake, N., & Koschmann, T. D. (2002). Realizations of CSCL conversations: Technology transfer and the CSILE project. In T. Koschmann, R. Hall, & N. Miyake (Eds.), *CSCL 2: Carrying forward the conversation* (pp. 3-10). Mahwah, NJ: Lawrence Erlbaum.

O Burton, G. (1996-2004). *Silva Rhetoricae*. Brigham Young University. Retrieved August 16, 2005, from http://humanities.byu.edu/rhetoric/Persuasive%20Appeals/Ethos.htm

O'Malley, C. (Ed.). (1995). *Designing computer support for collaborative learning*. In *Computer supported collaborative learning* (pp. 283-297). Berlin: Springer Verlag.

Ó Murchú, D., & Sorensen, E. K. (2004). Online master communities of practice: Collaborative learning in an intercultural perspective. *European Journal of Open and Distance Learning, 2004/I*. Retrieved from http://www.eurodl.org/

The Oxford Encyclopedic English Dictionary (1991).

Pea, R. (1996). Seeing what we build together: Distributed multimedia learning environments for transformative communications. In T. Koschmann (Ed.), *CSCL: Theory and practice of an emerging paradigm* (pp. 171-186). Mahwah, NJ: Lawrence Erlbaum Associates.

Roberts, T. (Ed.). (2004). Preface. *Online collaborative learning: Theory and practice* (pp. vi-xii). Hershey, PA: Information Science Publishing.

Salmon, G. (2000). *E-moderating. The key to teaching and learning online.* London: Kogan Page.

Scardamalia, M., & Bereiter, C. (1996). Computer support for knowledge building communities. In T. D. Koschmann, R. Hall, & N. Miyake (Eds.), *CSCL theory and practice of an emerging paradigm* (pp. 249-268). Mahwah, NJ: Lawrence Erlbaum Associates Publishers.

Sorensen, E. K. (2004). Reflection and intellectual amplification in online communities of collaborative learning. In T. S. Roberts (Ed.), *Online collaborative learning: Theory and practice* (pp. 242-261). Hershey, PA: Information Science Publishing.

Sorensen, E. K., & Ó Murchú, D. (2004). Designing online learning communities of practice: A democratic perspective. *Journal of Educational Multimedia (CJEM), 29*(3), 18.

Sorensen, E., & Ó Murchú, D. (2005, September 20-23). Developing the architecture of online learning communities: Designing the walls of the learning space. *Proceedings of the 11th Cambridge International Conference on Open and Distance Learning: The Future of Open and Distance Learning.* Cambridge: Madingly Hall.

Sorensen, E. K., & Takle, E. S. (2002). Collaborative knowledge building in Web-based learning: Assessing the quality of dialogue. *The International Journal on E-Learning (IJEL), 1*(1), 28-32.

Sorensen, E. K., & Takle, G. S. (2004). Diagnosing quality of knowledge building dialogue in online learning communities of practice. *World Conference on Educational Multimedia, Hypermedia and Telecommunications* (Vol. 1, pp. 2739-2745).

Tait, A. (2003). Managing student support in distance education. In S. Panda (Ed.), *The management of distance reducation.* London: Kogan Page.

Wenger, E. (1998). *Communities of practice: Learning, meaning and identity.* Cambridge, UK: Cambridge University Press.

Wenger, E., McDermott, R., & Snyder, W. (2002). *Cultivating communities of practice. A guide to managing knowledge.* Cambridge, MA: Harvard Business School Press.

Wilson, C. (1998). *The books of my life*. Charlottesville: Hampton Roads Publishing Company.

Elsebeth Korsgaard Sorensen, Aalborg University, Denmark

Daithí Ó Murchú, Gaelscoil Ó Doghair & Innovative e-Learning/e-Tutoring, Hibernia College, Ireland

February 2006

Acknowledgments

We owe a great deal of gratitude to people who, in various ways, helped to make this book possible. We would particularly like to thank all of the authors without whose committed investments and willingness, in each their individual ways, to "go beyond the box," this book could not have come into existence. We also thank everyone who participated in the review process. In this respect, special gratitude should be given to Knut Steiner Engelsen, Art Ó Súilleabháin, Sandra Weinreb, Andrew Goh, David Topps, and Tim Savage.

We also like to mention the Cambridge International Conferences as a rich source of inspiration. Of special note is Alan Tait, Professor of Distance Education and Development at the Open University, UK, who invited and supported the idea while it was still in its cradle.

Although much of our collaborative work on this book was carried out in cyberspace, beautiful spots on the wonderful coasts of both Denmark and Ireland served as peaceful and reflective working venues for each of us. A special thanks to Marianne Kollander and Morten Noreng for contributing to such working conditions on the Danish side, and every special person who gave space to reflect and write in Ireland.

Thanks also to the Faculty of Humanities, Department of Communication, for providing brilliant opportunities for work and research online, and to the Gaelscoil family and colleagues at Hibernia College.

Finally, we would like to thank our spouses, Daniel and Norita, and our children, Rune, Jon, Maiken, Céadlyn, Bronna, Suan, and Laoilan, for their patience and love in stressful times.

Elsebeth Korsgaard Sorensen, Aalborg University, Denmark

Daithí Ó Murchú, Gaelscoil Ó Doghair & Innovative e-Learning/e-Tutoring, Hibernia College, Ireland

February 2006

Chapter I

Online Communities and Professional Teacher Learning:
Affordances and Challenges

Norbert Pachler, Institute of Education, University of London, UK

Caroline Daly, Institute of Education, University of London, UK

Abstract

This chapter examines online communities for professional teacher learning, in the context of a mixed-mode practice-based Masters degree, the Master of Teaching. It problematises key principles for designing collaborative learning online, aimed at developing teachers' dispositions and values, as well as critical understandings that inform professional knowledge about practice. Data from teachers' asynchronous online discussions are analysed, and the discussion is grounded in the learning activities of course participants. The authors establish the contemporary context for developing teachers' professional learning through the affordances of new technologies, with a view to establishing what claims can be made about the potential of online communities to provide a counter to reductive models of professional development that have dominated teachers' learning in England and Wales in recent years.

Introduction

This chapter is grounded in research and development work in the area of professional teacher learning on an award-bearing programme conducted by 10 plus higher education tutors, and some 100 plus master's level students. The sample of primary data presented here draws on asynchronous online discussions that took place within a tutor group of 15 teacher participants, over a total 9 month period, as part of the programme. Grounded in actual practice, the chapter discusses a number of design principles for online communities for professional teacher learning, with a view to sketching out the beginnings of a conceptual and theoretical framework for the area. Invariably, in view of the limited space available, some key issues and notions are being dealt with rather cursorily, and only a comparatively small snapshot of primary data can be presented and analysed. Readers are encouraged to engage with a range of other outputs by the authors, where some of these issues are discussed in more depth (see references). The main body of the chapter draws on the data from three online discussions, each lasting 4 to 5 weeks, that took place within a password-protected forum. Participants are identified here by the first letter of their forenames (for example, Teacher K), and they form an online tutor group of 15 primary, secondary, and special school teachers with between 2 to 5 years teaching experience. Their writings have contributed to our refinement of the design principles for learning within online discursive communities that are set out below. Contemporary contextual factors, which have informed our conceptualisation of teacher learning in this environment, are presented at the beginning of the chapter.

New Technologies and Professional Teacher Learning: The Context

The role of new technologies in professional teacher learning is an area that is as yet scarcely explored, despite the huge expansion of these technologies in schools, and the education system in general, over the past decade. Teachers in England and Wales now work routinely with electronic media in carrying out a range of professional duties, both managerial and pedagogic. While we do not subscribe to the conventional dystopia, dystopia polarisations, the transformation of the school into a workplace supported by new technology has both huge potential for teacher learning and the transformation of teacher roles. But it also carries with it the danger of teachers being subsumed into the reductive discourse of efficiency and effectiveness, within a political agenda driven by an economic rationalist view of education as performance-driven. In England and Wales, teachers often see themselves occupying bureaucratic and managerialist operational roles in implementing centralist, target-driven agendas, supported by the increasing availability of data manipulation via technology. What is distinctly lacking in the expansion of technologies into the professional domain, is a review of what they have to offer as pedagogical tools for professional learning within a broader view of what the aims of teacher and pupil learning might be, and, thus, of the type of profession and education system we want to see emerge. It is impossible to enter into debate about new technologies and teacher learning without having an ethical orientation towards the *purposes* of technology in relation to the *purposes* of education.

Thus, we see a chief potential of new technologies to be in offering a counter to the reductivist discourse of teaching and teacher education that has dominated for the past decade, and that has been assessed by Wrigley (2004) to have narrowed the purposes of continuing professional development (CPD), along with narrowing the purposes of education at large. As part of the bureaucratic/managerialist discourse, technology has played a central role through the proliferation of websites–government, local authority, and commercial–that disseminate centralised policy-making, coordinate the standardisation of pupil and school performance by statistical analyses, and provide standard proformas for teaching and curricular content. Teachers can now download entire corpora of content information for teaching, minimising the autonomous and critical-thinking dimensions of their role. Teachers have, to an extent, been positioned as passive recipients of information about practice-based knowledge. Wrigley asserts that dissident voices now, however, have an "international strength" based on the numbers of teachers leaving the profession, and the concerns of those who manage schools. For the first time, there is a "genuine opportunity, despite the global drive towards 'effectiveness' in 'economic rationalist' terms, to build a movement for real improvement in education" (Wrigley, 2004, p. 242). The teachers themselves, and how they are enabled to learn as professionals, must be at the core of this new optimism concerning pupils' educational experiences.

The potential of technology to support collaborative learning based on notions of shared knowledge construction, and to build forums for the coconstruction of knowledge, has scarcely touched classrooms, and it is no surprise that it has touched teachers' professional development still less. This we see to be characteristic of how current political contexts have coincided with an early phase in the sociocultural evolution of technologies, characterised by Andriessen, Baker, and Suthers (2003) as a transition period from the "information age" to the "knowledge age." They explain that transition from individual, primarily content-based learning practices, to collaborative, knowledge-making ones, is an "uncertain process," and that inevitably, there exists a wide range of inconsistent views on how we can learn in this medium (Andriessen et al., 2003, p. 2). Within this uncertainty, however, lies the possibility of bringing about real change through innovation. At this stage in the development of computer-supported collaborative learning (CSCL) environments, for example, participants are part of a very recent, and constantly accelerating history of change in how learning can be organised and conceptualised. This shift away from the transmission-focused "information age" denotes an altered perception of people as having the capacity for agency, who share corporate responsibility for making their knowledge through collaborative processes. For teachers, CSCL can provide an alternative professional learning experience that involves the forging of a community of learners with altered roles, both in terms of their own learning and the learning of others. The challenge is to innovate now to shape this technological evolution in the domain of teachers' learning, and to resist the managerialist adoption of technologies into an education system absorbed in reductionism. Those of us who are involved in developing online courses for teachers have a vital opportunity to design learning experiences that are based on harnessing new technologies to help shape future professional learning and professional identities.

Research and curriculum development work undertaken at the Institute of Education, University of London in developing the accredited, mixed-mode Master of Teaching (MTeach; see http://www.ioe.ac.uk/mteach/), which provides the context for this chapter, foregrounds pedagogical principles based on collaborative theories of learning, which we would see in turn as impacting on teachers' understanding of how their own classrooms function, with or without technology. This requires a considerable epistemological shift in teachers in reappraising how pupils and teachers learn, and the role of technology in supporting this. It makes paramount the ethical responsibilities of teacher educators in designing for e-learning, so that it contributes to teacher learning that is "an increase in the skills and knowledge base as well as the understanding of teachers in relation to general pedagogical and didactic principles ...through mediated but self-directed engagement with peers, tutors and content" (Pachler & Daly, 2004). In this chapter, we examine the theoretical foundations on which the MTeach course is based, and present a model for coconstructed learning that implies a reassessment of teachers' professional identities with activist capacities, that is, a rich conceptualisation of the purposes of education and of how learning is brought about, both of which have consequences for the ethical aspects of the use of technology in a context of professional development.

The MTeach is a mixed-mode higher degree programme that accredits high-level professional learning. It is characterised by a focus on collaborative enquiry as the basis for initiating change, and achieving agency through practice-based learning, as well as developing educational and research literacy as part of the teacher knowledge base. The course is based on shared knowledge construction at a distance through computer-mediated communication supported by some face-to-face meetings:

In essence, the (MTeach) requires participants to provide public accounts of several aspects of teaching — e.g., classroom interactions (incl. pupil/ teacher talk, teacher questioning, grouping), the management of learning environments, the active engagement of learners, the design of materials, the evaluation of outcomes, the assessment of learning, the provision of feedback, the teaching of learner strategies, etc. — and learning — e.g., understanding personal and professional learning, understanding and evaluating pupil learning, leading and managing change — all of which receive critical review from peers and the tutor. Much effort is expended on asking and refining probing questions and pursuing investigations and enquiries. This is done in order to provide the basis for productive professional conversations. (Pachler & Pickering, 2003, p. 39)

Online Professional Communities: "Doing" and "Knowing"

Our discussion is predicated upon a paradigm of learning, in our instance, professional teacher learning, as a process that, while taking place by individual learners confronting personal construct systems, is embedded in a social context, and develops through communication, collaboration, and/or guided activities, in particular experimental-participatory experimentation (for a detailed discussion see Pachler & Daly, 2004). From this flows the pedagogical imperative, in our view, to require learners to take an active part in the development of processes by which they come "to know" about teaching and learning in critically informed ways. We see them having an active part in defining "that which is to be learnt" in respect of professional practice. The processes of learning are based in the negotiation of meaning that takes place through peer discussion around professional events, and experiences that are narrated to the online forum, and reviewed in relation to online textual material. This is by no means to imply that the process involves moving towards a consensus of meaning, but rather that "knowing" takes place through negotiating the significance of meanings that are derived from engaging with various professional practices. When teachers offer accounts of their own and others' professional activity to the forum, interactive exchanges about them take place over time, shaped by the properties of the asynchronous environment. The online exchanges develop altered conceptualisations, based on the capacity of teacher-narration, when it can be revisited over time from different interpretive angles, to offer countertruths to the prevailing rationalist discourses of performativity and effectiveness. The participants offer interpretations of each other's professional experiences, like the one below, where Teacher S offers an analysis of Teacher K's account of the obstacles she has encountered to improving her practice:

K. commented that self-evaluation is difficult...possibly because the 'reflective dialogue' (she) speaks of is often of no value when pressured teachers feel insecure, overworked and defensive...possibly too because we are so 'tuned in' to the measuring of students, i.e., predicted grades, target grades...that we forget about the learning.

The impact of new technologies needs to be seen as happening in relation to learning in practice, as well as learning as a matter of individual cognition or shared understanding. Wenger's work (1998) on communities of practice has been highly influential in developing conceptualisations of the formation of shared professional understandings and practices, and how far such a community may be realised within an online constituency of teachers. A community of practice, drawing on the work of Wenger, will operate through the twin processes of *participation* and *reification* by which practitioners render professional phenomena meaningful, and thereby construct and sustain what it is to "be" a member of that practice. A "community of practice" creates and sustains itself, and impacts upon practice by establishing amongst its members a shared repertoire, mutual engagement, and joint enterprise. The emphasis on practice informed by negotiation is particularly relevant to the context of teachers in England and Wales:

Practice...exists because people are engaged in actions whose meanings they negotiate with one another.... (It) resides in a community of people and the relations of mutual engagement by which they can do *whatever they do...(our italics). (Wenger, 1998, p. 73)*

The effectiveness of computer-mediated communication (CMC) for teachers' professional learning, then, may be judged by how far it enables such engagement, the quality of the resulting negotiation, and whether this has an effect upon what teachers *can do* in the professional domain of the classroom. For a community of practice, however, this goes beyond a functional or delivery mode in terms of what teachers can do: instead it places "doing" in a context of enhanced abilities to make judgements and to be discriminating in the light of shared professional experiences:

Becoming good at something involves developing specialized sensitivities, an aesthetic sense, and refined perceptions that are brought to bear on making judgements about the qualities of a product or an action. That these become shared in a community of practice is what allows participants to negotiate the appropriateness of what they do. (Wenger, 1998, p. 81)

The development of discriminating, qualitative judgements leading to action is core to how we design the online component of the MTeach. These judgements are not so much derived from individual cognition, though resulting actions may be (and most often are) individual, but are rooted in shared coconstructed knowledge — what it is "to know" about teaching and learning in order to "become good" at it. Participants' exchanges frequently refine the telling of "experiences" into negotiation for action:

Having [participated] in this discussion I have been trying to use the term 'learn' rather than 'work' when introducing the days activities to my class of brand new (summer born) reception [children]. We already made a good start in the nursery introducing ways in which we learn (through our senses in the most basic terms for 5 year olds) and this will hopefully build upon that. (Teacher S)

In response to [Teacher D] I agree that the support and development [of teachers] seems to stop [after the NQT year]...department meetings should be much more philosophical...about talking and developing teaching and learning strategies and looking at examples of good educational research. (Teacher L)

In an interesting paper on the question of how people know, Deanna Kuhn (2001) argues that people's epistemologies, that is, "what they take it to mean to know something" (Kuhn, 2001, p. 1), influence how they make use of their intellectual skills, as well as the acquisition of new knowledge. She holds that this is true not only for knowing how one knows something, but also for how knowing processes operate (Kuhn, 2001, p. 7). She demonstrates the importance of the relative strengths of theoretical explanation vs. evidence as justification for causal claims, and she argues for the importance of helping people gain an understanding of the epistemic strengths and weaknesses of each. Kuhn sees both as complementary, namely as offering understanding vs. truth, and holds that, which is the better depends on the function that they are called to perform upon in the context of an argument. People need much more than the procedural skills to acquire new knowledge, since "the locus of individual differences in intellectual performance may lie at a less obvious level than that of the performance itself" (Kuhn, 2001, p. 8). Dispositions, she further asserts, are not dictated by cognitive competence and that therefore, motivation, and an appreciation of the benefits of knowing, are required to get people to make the effort necessary to acquire new knowledge. Finally, she sees people acquire these values and dispositions in social settings and not in isolation.

By way of an overview, Kuhn (2001, p. 4) offers an — in our view — persuasive diagrammatical representation of her line of argument, which we reprint here, with permission, in Figure 1.

What then Flows from These Insights, Assuming they are Borne Out More Widely, for Computer-Mediated Professional Teacher Learning?

Most importantly, we take Kuhn's findings to lend strong support to a conceptual design framework, and to pedagogical practice that foregrounds a constructivist and social-interactivist paradigm based on notions of inquiry, analysis, inference, and argumentation.

And, equally importantly, we want to stress the dispositional and values orientation of Kuhn's model that, transposed to the context of professional teacher learning, we interpret as grounded in teacher agency and criticality. Those processes by which dispositional orientations develop are important to understand in the context of CMC, where they play a major role in achieving ethical purposes in teachers' CPD. We believe that metalearning is central to the development of such dispositions, and that the links between thinking and

Figure 1. Metalevel competence and dispositional factors as contributors to intellectual performance (Source: Kuhn, 2001)

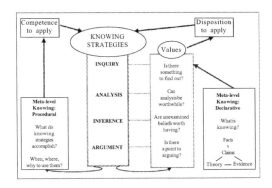

practice are secured in online Masters level discussion through a rigorous engagement with the negotiating of meanings in practice with peers, stimulated by questions and readings that force participants to address themselves as learners, and their relations with external agendas that affect what happens in their classrooms. Teachers L and M, are developing ethical orientations based on a shared understanding of appropriate practice:

The sad thing is that I often feel I do not teach very well because the other aspects of the job (initiatives from government, examination boards or senior management) receive more of my time and effort than my classroom teaching. We are torn between doing the jobs we have to do/are told to do and trying out the things you would actually like to with young people...[what] makes our job as a teacher interesting is...the human aspect...[when] the pupils realise the complexities of life, they too are self assessing and analysing. (L)

As [the research] discovered, it is impossible to consider students' learning without having to reflect on your own teaching. I maintain that teaching and learning go 'hand in hand' and would suggest that those finding it difficult to self-evaluate...may adopt the same principles they use when evaluating the work of their students. (M)

In order to develop values-based discussion, in view of Kuhn's findings, we deem it to be incumbent upon the designers of educational experiences, based on notions of knowledge generation, to take due account of metalevel engagement of learners, both at a procedural, as well as at a declarative level. We have developed four design principles, drawing on a range of theories that support metalevel engagement, in developing online communities for teacher learning:

- the narrative principle,
- the discourse principle,
- the argumentation principle, and
- the intercultural principle.

The Narrative Principle

A key aspect of metalearning in the MTeach has been the design of learning templates around teacher narratives. Conceptualisations of narrative draw on a range of ways of "knowing," from narrower views of narrative as primarily concerned with *form*, in which events, characters and ideas are organised according to certain stylistic conventions, to those based on *function,* which require a different understanding of narrative as involved in the production of identities, for example as defined by Polkinghorne (1988, p. 1):

Narrative meaning is a cognitive process that organizes human experiences into temporarily meaningful episodes. Because this is a cognitive process, a mental operation, narrative meaning is not an 'object' available to direct observation. However, the individual stories and histories that emerge in the creation of human narratives are available for direct observation. Examples of narrative include...the everyday stories we use to explain our own and other's actions.

Narrative, in these terms, has informed the view of teacher narrative as arising from, and constituting "critical episodes" (Tripp, 1993), and having a vital role in enabling teachers to challenge orthodox, universal "truths" that govern their practice (Clandinin & Connelly, 2000). Here, narrative *function* has been seen to contribute to teachers' learning by the ways in which (professional) subjectivities are constructed through the "telling" of experience from professional life. Our interest lies in the "verisimilitude" of such narratives of experience, which Bruner argues are constructed by individuals to organise their experiences in order to make sense of them (Bruner, 1985). This function of narrative we see as supported by peer interaction online over time. CMC in the MTeach helps participants to develop deeper, coconstructed, professional meanings, based on a range of narratives that they tell online, in which they:

- narrate critical episodes from their classrooms,
- explain their dispositions towards educational research, based on their experience,
- give accounts of particular teacher-student interactions,
- describe a moment from their own learning,
- give a rationale for a choice of topic for a professional journal, and
- coanalyse a classroom episode that has been peer-observed.

Such experiences form the basis for the range of online discussions that are conducted within the password-protected MTeach forums, in tutor groups of up to 15 participants. The shared narrative approach encourages the construction of meaning arising from those narratives, forged with peers, over time, through the asynchronous environment. An example of a teacher narrative, which triggered discussion around the ethical and practice-based implications of "initiatives," came from Teacher S:

At our INSET (In-service Training) day this week, our literacy co-ordinator talked about research carried out on boys and reading levels. She provided the staff with quotes, data etc. to back up several ideas put forward to develop boys' reading levels. Yet when I went to talk to her afterwards she admitted that she didn't believe everything she'd just told us but felt we needed to be informed. Had she been research literate she would have been able to make much more use of this research.... (S)

This teacher's narrative played a provocative role in the discussion about the role of research for teachers, as much as the "authorised," online, academic texts we provided.

Textuality

Earlier, we alluded to the importance of textual interaction in computer-mediated communication, and it is in the affordances of shared textual authorship that we believe lies the most potential for deepening thinking over time, and enhancing conceptual transformations for online participants. Blommaert, Creve, and Willaert's (2004) assertion of the "normalcy" of textual literacy in the defining of what it is to be "normal members of our cultures," has particular application to the electronic environment, where being readers and writers of electronic text, written within the forum, establishes a mutuality, or what may be termed "moral purpose" (Fullan, 2001) among the community of fellow participants.

Notions of literacy change over time, inter alia, to accommodate increasingly broader conceptualisations of text. Also, definitions of literacy inadvertently comprise narrative in view of the prevalence of stories of all kinds in everyday life. Narrative can be of considerable influence on perception and belief (Kuhn, 2001, p. 1). As Robertson, Shortis, Todman, John, and Dale (2005) point out, the use of new technologies for learning, amongst many other things, is characterised by a destabilisation and decentring of existing "orderings" of classroom practices, often leading to ruptures in the flow of pedagogical practices. In addition, it is accompanied by its own "technical-regulative and instructional-pedagogic discourse," as well as new pedagogic identities for teachers and learners. This perspective on pedagogic discourse highlights the view that engagement in electronic, text-based discussion may exhibit "the potential of disordering" which according to Bernstein (1996, p. 128), is the inevitable coeffect of attempts to establish "a modality of order" by which conceptual space may be governed. ICT can be seen to recontextualise the subject discourse, as well as impose an ICT procedure discourse. ICT brings with it a certain unpredictability, and invariably, adds to the complexity of learning environments.

Luckin, Plowman, Laurillard, Stratford, Taylor, and Corben (2001), for example, show that narrative can help the learner overcome the fragmentation of thought implied in hypertextual structure. Although other textual genres offer alternative organising structures for thought, their argument is that narrative "is fundamentally linked to cognition and understanding" (Luckin et al., 2001, p. 100). They acknowledge the diversity of perspectives on narrative, including in relation to other means of textual organisation and outside of the online environment, and emphasise that they develop a "working definition" for narrative that can be applied to interactive educational media:

Narrative is a process of both discerning and imposing structured meanings which can be shared and articulated. The result of this process is also often referred to as a narrative, i.e., the product of discerning and imposing structured meanings which can be shared and articulated. (Luckin et al., 2001, p. 101)

Different ICT artefacts make differing demands in terms of potential capabilities. Hyperlinked and distributed digital-learning material and objects are often characterised by nonlinearity, and require of the learner goal-directed, active application of narrative principles according to prevailing narrative modality and perspective.

Different approaches to narrative and learning can be discerned, each positing a particular definition of the concept, according different roles to the recipient and the narrator, and implying different roles for technology. For our purposes on the MTeach, the notion of metanarrative as a macrostructure, that is, the ability to make individual postings of reflective texts, based largely on professional practice by course participants, cohere and connect into some form of narrative is key. In Laurillard's terms:

Narrative provides a macro-structure, which creates global coherence, contributes to local coherence and aids recall through its network of causal links and sign posting. The structure provides a linear dynamic using a variety of structural cues, such as headings, textual signposts, and paragraphing, to allow learners to maintain their plans and goals. It has both cognitive and affective impact, performing an essential organising function for the learner by shaping the creation of meaning from texts of all kinds. (Laurillard, Stratfold, Luckin, Plowman, & Taylor, 2000, p. 2)

These properties contained within narrative macrostructure are, of course, not limited to the online context. What is distinctive about the intervention of asynchronous CMC, in this context, is the effects of copresence and coauthorship of text on the construction of the macrostructure over time. A longer extract from a discussion on "What is good educational research?" reflects the participants' copresence in a text authored by one teacher, J, but constituted by the collective thinking, from which it has evolved:

I work in an "SEN" environment and am surrounded by professionals with a wide range of expertise...The sharing of ideas relating to teaching and learning strategies is often disseminated via spontaneous conversations during a rushed lunch break...As [Teacher L] considered, "should our environment become more philosophical?"

I believe that until such a time when real importance is given to listening to those who are involved in the job, at the frontline, then there will never be a place for sharing examples of educational research — whether good or bad!...[Teacher G] uses the term "lazy", suggesting that many who oppose change do so through fear of work. There are probably a small minority of teachers who may fit into [Teacher G]'s category. However...I believe many teachers have become despondent and negative towards change because it can be a genuine human fear.... How often are we as teachers told or given the research to back up why change *is occurring, or a* new *initiative being introduced? If teachers were more involved in the research that underpins such* changes *do you think they may be more receptive? Would this then be good research?*

There are extremely effective teachers who work within their classrooms often "blissfully" unaware of what is going on outside of their four walls. There are the teachers who are aware of what is going on all around them and have a constant desire to find out more, become involved and have an input or impact on teaching and learning. Considering [Teacher M]'s questions, "is it helpful to teacher's professional development if there is good research?" and "What do we mean by development?" I suggest that we need to accept difference, acknowledge and respect experience and accept that what is professional development for one may be completely different for another...[Teacher P] made the comment that the value of research "may be in how teachers view their roles within the school system," and he goes on to say this may have an indirect positive effect on classroom practice. I agree with [Teacher P] and would broaden his comment by saying that the value should be in how teachers view their role within the education *system and this could have an indirect positive effect on* teaching and learning *within the classroom and education system as a whole.*

Interdependence is accentuated by the oscillation by participants between the electronic forum and their continuing everyday professional life. Online textual production is conditioned by having to think with others in mind, others with whom one shares a textual, as well as professional, environment that is capable of continuous modification and reworking via the speed, accessibility, and transmutability of electronic text. The electronic environment records thoughts in writing as "stuff to be reworked" (as with conventional writing), but offers immediate synoptic capacities for self-review within the context of the peer-text, or macronarrative, which is constantly evolving. The "others" that are kept in mind have a particular function in modifying thinking that has much in common with, but also departs from the effects of those for whom we write as an "audience" in conventional terms, or those with whom we discuss in spoken discourse in say, a face-to-face seminar context. They are at once a constant, critical backdrop and active participatory audience. The online presence is constituted by individuals with their own patterns of cognition who, at the same time, contribute to the coconstruction of meanings that are forged through the requirement to write to others online.

If one accepts constructivist and social-interactionist notions of learning, whereby knowledge is developed through our interaction with the world around us, as well as through the process of rendering our everyday experiences of life meaningful, micronarrative, in the sense of storying, and metanarrative, as defined previously, have an important role to play in learning, particularly in environments without physical teacher presence.

Luckin et al. (2001, p. 119) posit an interesting distinction between what they call "designed-in" narrative, and the narrative as perceived by learners. The latter throws up interesting questions, for example, about the extent to which narrative structures built in by the material designers are productive, and to what extent they are actually realised and drawn on by learners. Learning templates on the MTeach have "designed-in" narrative as an explicit learning device. For example, the starter online task, set for newly qualified teachers (NQTs) who join the course as they begin their first teaching posts, is geared away from "survivalist" discourses, to establish a course principle of embedding teacher learning in the shared negotiation of practice over time, facilitated by the asynchronous environment (see Figure 2).

Figure 2. Starter task "Understanding Teaching"

Compose and share approximately 200 words on one of the following:

1. Describe a teaching group that you are working hard to "get on top of."
This may take the form of a short narrative.

OR   2. Describe a pupil who mystifies you, but for whom you would really like
to do something.

Remember not to use people's real names or write in a way that identifies individuals. You may choose an experience that you feel is particularly challenging -- possibly something on which you would like to hear some views or reactions -- that may help you make sense of the situation you describe, and help you plan future actions. You should finish your piece with a careful question that may help draw out helpful and practical responses from your fellow participants and your tutor.

Figure 3. Response partner task "Leading Learning"

Choose the focus for the journal that you will keep next half-term (this might be in particular lessons, working with other adults, mentoring, or leading an initiative, and so forth). Write a rationale for your choice of focus ... . This should give a context for your plans to develop learning (your own/your pupils'/your colleagues'). Consider

What reflections can you make about your experience of leading learning, or of being led?

What are your personal reasons for your choice, and what extrinsic factors affect it?

How far is your focus something that you think can contribute to the learning of others, pupils, and teachers?

Response task

Read the task posted by your response partner. Write a brief response (150-300 words) to the rationale, including asking one or two questions that you think could help your partner to focus his or her reflections.

Linked to the conceptualisation of narrative is the notion of affordability of digital learning objects and (textual) artefacts; or, put more pointedly, what pedagogical features that are most conducive to learning are embedded in the technical design? Technology can also be used to produce narratives in the sense of stories for learners to work with, for example, to develop memorisation, understanding, and/or problem solving. On the MTeach, participation in online discussion is a compulsory aspect of the mixed-mode programme, and is essential to the establishment of mutuality on which the ideas of professional learning are built. The discussion itself provides a coconstructed rendering of teachers' experiences, as they negotiate a range of professional practice that goes beyond the "designed-in" narrative of the course design. It is at this stage that teacher action is explored with a depth of critically informed reflection, and it is here that the most potential for impact upon practice has emerged, as teachers have adopted counterarguments to the prevailing political discourses in which they work, and begin to voice counternarratives that are based on close examination of the recount of practice:

I do not agree (as it would seem most of the group would not) with [Teacher O] who does not think it necessary for teachers to be research literate. If

this was the case, teachers would teach anything presented to them without question.... I believe by the nature of our job we have the ability to think critically a range of 'research-based' initiatives presented to us. Through our own practice we are able to identify weaknesses and make judgements. The research commented on by [Teacher S], regarding the underachievement of boys, we all have our own response to according to our experiences. [Teacher J] says that if we are research literate we can make decisions for ourselves, rather than letting others do this for us. (Teacher G)

The Discourse Principle

In the context of computer-mediated communication, questions arise around the issue of how best to enable participants to make the necessary links, for example, through the use of threading facilities, and so forth. Rather than looking solely to technical solutions, which can be seen as problematical, for example, in terms of the difficulty to represent conceptual, rather than historical development (Turoff, Hiltz, Bieber, Fjermestad, & Rana, 1999), in relation to the creation of the necessary links between postings, we would stress the importance of the pedagogical design underpinning the learning process. Turoff et al. (1999) propose the development of what they call "conceptual discourse templates" or "domain independent general metadiscussion structures" to facilitate group metacommunication processes, and to ensure individual participants can capture and categorise the majority of a discussion in large online groups. They hold that a transaction analysis view around discourse models, conceived of as speech act and syntactic grammar models is less fruitful compared with discourse models "which allow individuals to classify their contribution to the discussion into meaningful categories that structure their relevance and significance according to the nature of the topic, the objective of the discussion, and the characteristics of the group." This, they posit, would allow best to account for the complex interplay of temporal occurrence, comment/reply hierarchy and keyword association of comments in discourse structures. In addition, they argue for conceptual diagrams to lessen the cognitive burden caused by a lack of spatial and temporal context to be built collaboratively within a group, giving individuals the opportunity to modify concept diagrams to facilitate better understanding, and groups of participants a visual representation of the shared understandings (see also Pachler & Daly, 2003).

To this end, we have developed the following bespoke, yet very simple, pedagogical framework on the MTeach: online participation adheres to a task structure that we view as a "learning template" encouraging participants to "go meta" about their teaching. We have become increasingly aware of the need to vary the template described here in order to avoid it becoming "regulatory" and restrictive of negotiation of meaning among course participants. Participants are involved in a series of online discussions, each of which follows a similar pattern:

...an opening page delineates briefly the aims, purpose and context of the discussion within the module in which it is located. From this, participants move either to the task itself or to a background paper written by course tutors drawing on key literature in the field and listing selected background reading. Having read the background paper and engaged with the related digitised readings participants post their response by email to their online tutor group. They further submit at least one response per online discussion to their peers' postings, after which tutors summarise participants' contributions and, thereby, "close" the discussion.... (Daly & Pachler, 2004)

This model broadly adheres to Laurillard's notion of the learning process being akin to a "conversation" between the teacher and the student (Laurillard et al.,1993) that operates at a discursive and interactive level linked by reflection and adaptation (see Figure 4 from Laurillard et al., 2000).

Laurillard's model is based on Pask's (1976) conversation theory in which learning is seen in terms of conversations between different systems of knowledge. According to Naismith's (Naismith, Lonsdale, Vavoula, & Sharples, 2004, p. 15) interpretation of Pask's model, in order to learn, a person must be able to converse with him-/herself about what s/he knows and, in order to learn more effectively, converse with other learners by interrogating and sharing their description of the world. Within a text-based electronic learning environment, language-based models of learning are central to the design for such "conversing" to occur.

Learning is a continual conversation: with the external world and its artefacts, with oneself, and also with other learners and teachers. The most successful learning comes when the learner is in control of the activity, able to test ideas by

Figure 4. The conversational framework for the learning process (Source: Laurillard et al., 2000)

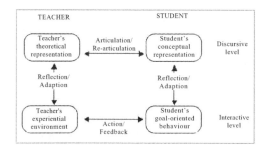

performing experiments, ask questions, collaborate with other people, seek out new knowledge, and plan new actions.

On the MTeach, we concur with the distinction of students and teacher operating at different levels, as well as the view that students adapt their actions toward "becoming good" at what they do in the light of reflection on their current conceptual understanding. We also agree with the importance of action and feedback, as well as articulation and rearticulation. However, unlike Laurillard, we do not see these processes mainly as an outcome of "conversations" between the student and the teacher but, as noted by Naismith et al., equally, if not more importantly, as an outcome of "conversations" among students. This perspective motivates our pedagogical approach, and the way the teachers experience the online environment:

One thing I wanted to bring to the attention of the group was that the online discussions have been almost like a journey for me...I start out believing in one thing and...I end up thinking another! Am I the only one this happens to? Hope not! (Teacher A)

In a further refinement of the conversational principles underpinning pedagogical online design, we build in the critical role played by texts in provoking metalevel discussion. (For a detailed discussion, see Pachler & Daly, 2004.) Relations between participants and the textual environment are instrumental in achieving critical levels of discussion, and are constantly negotiated between participants as they interpret their practice in response to online readings and peer-authored contributions. The model in Figure 5 represents three learning trajectories for

Figure 5. Online professional learning trajectories (Source: Pachler & Daly, 2004)

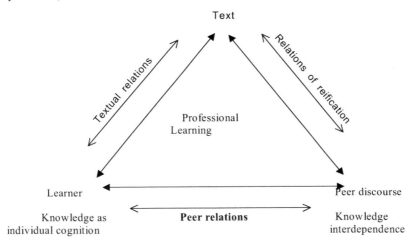

teachers that are discernible in the MTeach online environment, and that are constitutive of Wenger's concept of the *participation* and *reification* of professional phenomena. Within this model, the relations that exist between texts, the individual learner and peer discourse constitute the development of practice-based professional knowledge. Differing degrees of autonomy/interdependence exist along each learning trajectory, dependent upon the configuration between the sets of relations that operate within the online environment: textual relations, peer relations, and relations of reification.

- **Learner-text trajectory:** Learners are positioned according to how they view text-reader-author relations, how they view the teaching text as containing a unified, authoritative meaning, and the extent to which they seek to replicate "closed" texts in their online exchanges. The validity of text is reliant upon individual experience, and personal relevance. This trajectory falls within the individual-cognitive view of "intentional learning."

- **Learner-peer discourse trajectory:** Learners are positioned by the peer relations within the group, and their own orientation to knowledge acquisition and interdependence through interaction. This trajectory falls within social-constructivist learning perspectives.

- **Peer discourse-text trajectory:** Learners are positioned by the extent to which they participate in group discourse which engages in processes of reification of professional phenomena, by collaboration in meaning-making in response to text. This learning takes place through the forging of shared professional belief systems and discourses of practice. It falls within a participatory learning perspective, related to the development of corporate professional identity and agency (Pachler & Daly, 2004).

The relations we propose are shaped by use, and have something in common with what Goodfellow (2004, p. 387) has called "subjective recognition," which exists within the "operational," "cultural," and "critical" engagements of online participants. In the context of professional learning, such engagements go beyond facilitating either individual or shared cognition, or even practice, to become processes of professional making. Such processes are not dependent on complex uses of technology, but rather a design that capitalises on conversational models of learning. We would argue for a rather low-tech interpretation of the role of technology in the context of professional teacher learning, whereby it is mainly viewed and used as a tool enabling fellow learners to converse with each other and, thereby, engage in the processes of reflection, adaptation, articulation, and rearticulation, guided and supported by the teacher, rather than for a systems-approach, whereby the technology takes on some, or even many, of these core functions itself. (For a contrasting perspective on a technologically complex systems approach, see Naismith et al., 2004.)

The Argumentation Principle

One reason for espousing this view lies in our belief in the social embeddedness of knowledge creation. New technologies have a meaningful contribution to make to collaborative activity-, inquiry-, and discovery-oriented processes, which to us is central to knowledge creation: and, they facilitate a move away from content-orientation. Andriessen et al. (2003) view argumentation as an important realisation of collaborative activity, requiring learners to confront cognition. As cognitions can be seen as representations of the mind, they find articulation in points of view, and so forth, which in turn brings semiotic activity centre stage: semiotic, rather than more narrowly linguistic, as language is, but one semiotic system available for communication. They (Andriessen et al., 2003, p. 5) view argumentation as an epistemic activity, as for them, it involves the expression of knowledge and the relationships between pieces of knowledge. We concur with Andriessen et al. (2003, p. 4) in not viewing computer-supported collaborative learning environments as group learning situations without copresence, but rather "as groups of new integrated resources that can favour collaborative learning, using varied learning resources, structured interaction histories, interaction management tools and multiple representations of both problem-solving domain and interaction tasks." Inherent in the above is a perspective on learning as resolving different points of view by argumentation. With reference to Walton (1989, p. 10), they distinguish the following types of argumentation dialogue that, in our view, can all be meaningfully deployed in engaging learners in confronting cognition by producing and comparing arguments and using different types of reasoning and thereby learn

- quarrel,
- debate,
- persuasion (critical discussion),
- inquiry,
- negotiation,
- information-seeking,
- action seeking, and
- educational.

Andriessen implies that the configurations of "argumentation" are in fact, multiple. The structure of a "quarrel" online may be very different from dialogue that is "action-seeking," as they may be viewed as generically different forms, thus attracting different ways of organising ideas textually, according to the domain in which the participants are operating. Domains may be personal, political, aesthetic, as well as explicitly educational for example, and invoke different structures, such as antagonistic, knowledge-focused or consensus oriented. The social domains of knowledge construction involve differently appropriate language forms, here specifically textual structures within the particular frame of the written text. "Educational" may be more appropriately termed a "domain," rather than a specific type of argumentation dialogue: the educational domain could be said to embrace all of the argument forms listed above. Unsurprisingly, there is a distinctive sense of the educational domain that frames the generic forms of discussion engaged in on the MTeach. This typically generates forms of argumentation that are shaped by the professional context for the exchanges: they are committed to advancing knowledge; they are rarely "antagonistic," rather exhibiting a strong sense of collegiality and mutuality in building on what others have said; there is a shared participant identity, with frequent use of the collective pronouns "we" and "our" to describe both experience and thinking; they are particularly enquiring, and actively invite constructive peer-thinking to resolve professional quandaries, particularly in relation to taking action:

My questions for further discussion would be: How should I record these "narratives of experience" and my reflections upon them? How could they then be used to develop other teachers' knowledge and practice? Do they need to be shared with anyone else, or is it OK for them to influence my understanding alone? (Teacher K)

When referring to "narratives of experience," K. raised the question "Do they need to be shared with anyone else?" I personally think that initially "no," but once your knowledge and practice develops why not share that? As part of the "education" system is it not our role to educate each other as well as our students? (Teacher J)

In our experience, in order to maximise knowledge construction through online collaboration, a degree of "top-down" structure or proactive structuring of collaboration, that is, intervention in relation to how learning through discussion/argumentation should be organised, is necessary. On the MTeach, we structure learning by task, rather than communication tool-design. To some extent, we also structure collaboration "retroactively," that is, by regulating interactions through tutor facilitation. It is possible to extend the principle of argumentation into a highly regulated framework for the exchange of ideas, described by Jermann and Dillenbourg (2003, p. 205) as "scripts" that "formalize the educational context in which argumentation is expected to appear." MTeach discussions reflect what they might term a "scenario," in that there is a sequential organisation imposed on the discussion in which all participants are expected to act according to the conventions we set. This bears some relation to Jermann and Dillenbourg's scripted "phases":

A phase is descried by five attributes: the task that students have to perform, the composition of the group, the way the task is distributed within and among groups, the mode of interaction and the timing of the phase.

There can be seen to exist an interdependence between tool/learning environment and the activities used. On the MTeach, we believe that there are significant differences between learners in how they approach a discussion task. Whilst the template organises the phases of participation, it falls short of being a "script" in the sense intended here, as the development of dispositions is embedded in individual relationships between the teachers and the contexts of the classrooms within which they work, the textual material, and the collaborative textual environment. The relations referred to in Figure 5 have implicit in them the concept of wide differences in individuals' patterns of participation within the same tasks.

Andriessen et al. (2003, pp. 9-10) distinguish three types of learning in relation to "interactive argumentation produced in a normatively constrained and structured situation," that is debate:

- learning from the debate,
- learning about the debate, and
- learning to debate.

This, then, suggests that learners not only need to be acquainted and equipped with the requisite metasemiotic knowledge and conventions in order to part take in a meaningful manner in argumentative actions and interactions, but also that a certain prerequisite knowledge of the topic of a debate and associated concepts are important. Teachers possess the knowledge-in-practice to participate in such debates, but as well as developing critical understandings of that knowledge, the debate can produce the metalevel awareness that we have deemed vital to the ethical and dispositional aspects of teacher learning:

What have I learned about learning? That it is complex and you can't assume all learn the same, but that the main factor is the willingness of the person to learn. (Teacher L)

Also, Andriessen et al.'s (2003) distinction of the three types of learning suggests that there are benefits to be had from receptive, as well as productive engagement, also known in CMC-related terminology as "lurking."

All this poses the crucial question whether argumentation represents a viable option in bringing about knowledge construction in the context of the core theoretical principles underpinning the online pedagogy of the MTeach: collaboration, interdependence, teacher agency, and involved professionalism.

The Intercultural Principle

Intercultural learning contributes to teachers' learning-for-action and, in part, affects where they are placed in online learning trajectories (see Figure 5). It is rooted in the growing importance of the cultures inherent in online communities. With particular reference to identity, intercultural learning is embedded in the increase in cross-national, cross-cultural, cross-linguistic, cross-institutional discourse and learner interaction. In a recent review article in the field of second language acquisition/applied linguistics, Kern (Kern, Ware, & Warschauer, 2004) notes that intercultural learning and cross-cultural understanding did not automatically result from online communication. Nevertheless, cross-cultural online group composition can have a significant effect on the nature and extent of the learning that is taking place. The ability to bring together learners with different cultural backgrounds represents a significant new variable, for example, in terms of individual differences, educational histories, linguistic capabilities, cultural values and dispositions, and personal and professional experiences. In addition, the notion of "cultures-of-use," that is, the particularities or cultures of specific online groups and programmes is a formidable challenge not least as it frequently differs significantly from the educational cultures learners have been socialised into as well as across social, generational, institutional, and national groups. Goodfellow (2004, p. 387) argues that in the context of online distance learning environments for master's students, key institutional practices, which we would classify as "culture-of-use," are constructed from a mixture of differently mediated texts and procedures.

The operational, cultural, and critical engagements that students on…MA courses enter into with the university, their teachers and fellow students, are thus shaped not by any objective collective expression of identity or purpose, such as a building or set of occasions for gathering such as lectures, but by their subjective recognition of the meanings and values that these texts and practices enact. In the case of the majority of MA students, this recognition proceeds from prior experience of higher education and a professional background in education, but even...in this relatively homogeneous group, cultural and linguistic differences produce sharply different interpretations of the relations of power governing whose meanings and values predominate. Goodfellow (2004, p. 387)

In the context of the MTeach, this "recognition," and its attendant meanings and values, are partly affected by the particular school cultures, subject epistemologies and educational histories participants come out of, specifically in relation to pedagogical traditions. This has an important bearing on the types and nature of contributions towards knowledge coconstruction, reflected in their learning trajectories in their relations with texts and peer discourse in the online environment. Teacher N has reservations about the constructivist frameworks that underpin the discussions, and his contributions frequently emphasise his belief in individual expertise as a more meaningful professional goal for him:

I liked [Teacher D]'s point that teachers are naturally sceptical. I think it comes with experience! I find that the more experienced I become, the more I like my own ideas and opinions on what works. This doesn't mean that I'm not open to new ideas and other people's research though! I need to be able to see the point, see the facts and see the results.

Throughout the discussions, he usually elicits fewer responses than others, not because there is anything to suggest that others resist this orientation to learning, but rather because of the way it is manifested in the form of his contributions, which are usually written in less invitational tones than others, and stand as isolated treatments on the topics under discussion. Teacher R, although he works in the same school, has a very different perspective on the professional merits of collaboration, and takes a different stance towards shared professional learning:

Some people believe that recounting one's own experience of teaching is a highly personal thing and that there is no real need to share it with others...I am not convinced. I think that the more experiences that people can hear about the better. One thing that I learnt early on in my career was that if we all function as 'islands' then the profession is in danger of stagnating....

The continued coexistence of these differing stances would support Kern et al.'s assertion that online communication will not necessarily transform individuals' culturally embedded learning orientations. Another interesting consideration in relation to intercultural learning is the taking account of different orientations and values of participants or, in Hofstede's (1997, 2003) terms, "mental programs," which are largely determined by the social environments in which they were socialised. And, of course, the question of what cultural affordances learning technologies possess begs asking. Thorne (2003), for example, points out that communicative practices, whilst bound to the materiality of the medium, are not determined by it: rather they are negotiated dynamically through "cultures-of-use."

Conclusion

It is our belief that new technologies have the capacity to bring about fundamental and much-needed revision of the purposes of education, and what is considered appropriate professional development for teachers. Within a context of an online community of teachers, shared practice does not imply homogeneity. Rather, it affords the capacity for practical variance based on a reevaluation of how they act within the learning cultures that contribute to their professional making. In summary, this capacity involves:

- helping participants overcome some of the disadvantages of narrowly psychological perspectives of learning;
- rejecting attendant notions of teacher autonomy, that is, isolation, introspection, and self-referential practice;
- developing online teacher constituencies as communities of practice;
- engaging with narrative learning as both "designed-in" and emergent from online discussion;
- engaging teachers in the coconstruction of professional knowledge and metalearning;
- a reassessment of what it is "to know," within a broader reevaluation of the teacher's role in bringing about pupils' learning;
- committing to conversational principles as a core aspect of knowledge construction;

- a special role for texts as negotiating points in online environments; and
- emphasising qualitative analyses of the effectiveness of electronic learning environments.

Most critically, new technologies have a contribution to make in debating the professional and ethical principles that govern the choices teachers make about taking action that affects learning, both their pupils' and their own. Only from this perspective, can technologies affect the development of practice that is based on enhanced critical understandings of what they "can do," rather than what they are told to do. Never has it been more important to harness the true potentials of the technologies to work for genuine improvement, and to resist the transmission- and information-oriented, managerialist perceptions that have dominated the use of technologies in teachers' professional development to date.

References

Andriessen, J., Baker, M., & Suthers, D. (2003). Argumentation, computer support, and the educational contexts of confronting cognitions. In J. Andriessen, M. Baker, & D. Suthers (Eds.), *Arguing to learn: Confronting cognitions in computer-supported collaborative learning environments* (pp. 1-25). Dordrecht, The Netherlands: Kluwer Academic Publishers.

Bernstein, B. (1996). *Pedagogy, symbolic control and identity. Theory, research, critique.* London: Taylor & Francis.

Blommaert, J., Creve, L., & Willaert, E. (2004). *On being declared illiterate. Language-ideological disqualification in Dutch classes for immigrants in Belgium.* Paper presented for position of Chair in the School of Culture, Language and Communication at the Institute of Education, University of London, Spring 2005.

Bruner, J. (1985). Narrative and paradigmatic modes of thought. In E. Eisner (Ed.), *Learning and teaching the ways of knowing* (pp. 97-115). Chicago: University of Chicago Press.

Clandinin, D., & Connelly, F. (2000). *Narrative inquiry.* San Fransisco: Jossey-Bass.

Daly, C., & Pachler, N. (2004). Teacher learning: Towards a professional academy. *Teaching in Higher Education, 9*(1), 99-112.

Fullan, M. (2001). *Leading in a culture of change.* San Fransisco: Jossey-Bass.

Goodfellow, R. (2004). Online literacies and learning: Operational, cultural and critical dimensions. *Language and Education, 18*(5), 379-399.

Hofstede, G. (1997). *Culture and organizations.* New York: McGraw Hill.

Hofstede, G. (2003). *Culture's consequences. Comparing values, behaviors, institutions, and organizations across nations* (2nd ed.). London: Sage Publications.

Jermann, P., & Dillenbourg, P. (2003). Elaborating new arguments through a CSCL script. In J. Andriessen, M. Baker, & D. Suthers (Eds.), *Arguing to learn: Confronting cognitions in computer-supported collaborative learning environments* (pp. 205-226). Dordrecht, The Netherlands: Kluwer Academic Publishers.

Kern, R., Ware, P., & Warschauer, M. (2004).Crossing frontiers: New directions in online pedagogy and research. *Annual Review of Applied Linguistics, 24,* 243-260.

Kuhn, D. (2001). How do people know? *Psychological Science, 12*(1), 1-8.

Laurillard, D. (1993). *Rethinking university teaching: A framework for the effective use of educational technology.* London: Routledge.

Laurillard, D., Stratfold, M., Luckin, R., Plowman, L., & Taylor, J. (2000). Affordances for learning in a non-linear narrative medium. *Journal of Interactive Media in Education,* (2). Retrieved August 4, 2005, from http://www-jime.open.ac.uk/00/2

Luckin, R., Plowman, L., Laurillard, D., Stratford, M, Taylor, J., & Corben, S. (2001). Narrative evolution: Learning from students' talk about species variation. *International Journal of Artificial Intelligence in Education,* (12). Retrieved February 12, 2005, from http://aied.inf.ed.ac.uk/members01/archive/vol_12/luckin/paper.pdf

Naismith, L., Lonsdale, P., Vavoula, G., & Sharples, M. (2004). *Literature review in mobile technologies and learning* (Rep. 11). NESTA Futurelab. Retrieved February 12, 2005, from http://www.nestafuturelab.org/research/reviews/reviews_11_and12/11_01.htm

Pachler, N., & Daly, C. (2003). Computer-mediated communication and teachers' professional learning. Paper presented at the *British Educational Research Association Annual Conference*, Heriot-Watt University, Edinburgh.

Pachler, N., & Daly, C. (2004). *Professional teacher learning in virtual environments: Myth or reality?* Paper presented at the 11th International

Literacy & Education Research Network Conference on Learning, Havana, Cuba.

Pachler, N., & Pickering, J. (2003). Talking teaching: The Master of Teaching. *Change, 6*(2), 38-45.

Pask, A. (1976). *Conversation theory: Applications in education and epistemology.* Amsterdam: Elsevier.

Polkinghorne, D. (1988). *Narrative knowing and the human sciences.* New York: State University of New York Press.

Robertson, S., Shortis, T., Todman, N., John, P., & Dale, R. (2005). ICT in the classroom. The pedagogical challenge of respatialisation and reregulation. In M. Olssen (Ed.), *Culture and learning: Access and opportunity in the classroom.* Greenwich, CT: Information Age Publishing.

Thorne, S. (2003). Artifacts and cultures-of-use in intercultural communication. *Language Learning & Technology, 7*(2), 38-67.

Tripp, D. (1983). *Critical incidents in teaching.* London: Routledge.

Turoff, M., Hiltz, S., Bieber, M., Fjermestad, J., & Rana, A. (1999). Collaborative discourse structures in computer mediated group communications. *Journal of Computer Mediated Communication, 4*(4).

Walton, D. (1989). *Informal logic: A handbook for critical argumentation.* Cambridge: Cambridge University Press.

Wenger, E. (1998). *Communities of practice.* Cambridge: Cambridge University Press.

Wrigley, T. (2004). School effectiveness: The problem of reductionism. *British Educational Research Journal, 30*(2), 227-244.

Chapter II

Enhancing Phronesis:
Bridging Communities
Through Technology

Anders D. Olofsson, Umeå University, Sweden

J. Ola Lindberg, Mid Sweden University, Sweden

Abstract

In this chapter, the possibilities to use technology in order to improve the contextual and value-based dimensions in online distance-based teacher training in Sweden are explored. Aristotle's (1980) concept of phronesis is used as a starting point for raising questions whether the Internet, and the establishing of educational online learning communities, can be used to enhance the teacher trainees' skills of making moral decisions in unpredictable situations. It is argued that active participation, collaboration, and dialogue are vital in order to foster common moral and societal values among the teacher trainees, but that there is a need for rethinking how technology could be used in order to accommodate such processes. This chapter suggests that the development of a shared teacher identity is possible by expanding the scope of online community, and bridging teacher-training practices to teacher practices, thus including already practicing teachers, teacher trainers, and teacher trainees in a joint educational community.

Introduction

Society is changing. We have become part of a society characterised by multiplicity and globalisation. Teacher training is an institution buffeted by these changes. It must foster societal values such as democracy, freedom, multiculturalism, and equity, as well as ideas about teaching and learning, education, and instruction. However, can it do this within the ICT frameworks that have emerged in the learning society?

Enhancing learning through technology, in this chapter, is therefore conceptualised as a matter of being able to "walk the walk," in other words, of building bridges between the text-based learning environments of Web-based conference systems and situated practices relevant to teacher training. This can be done, we believe, by using an understanding of knowing in action based on the Aristotelian concept of phronesis, which captures the idea that engaged social practice — doing something "well," — has both contextual and moral dimensions (Aristotle, 1980; Gadamer, 1989).

In this chapter, then, we explore two issues. First, we comment on the work of an ICT-supported distance-based teacher training programme. Secondly, we consider whether online learning communities (OLC) are a valid basis for fostering a practice built around common societal values. Overall, our aim is to problematise the current relationship between online learning and teacher training. In short, can online "talk" be converted into classroom "walk"?

Fostering Values as Phronesis

Aristotle (1980) used phronesis to denote an aspect of knowledge that he claimed had both practical and moral implications, and he saw it as practiced rather than possessed, knowing rather than knowledge. Phronesis embraces prudence: the moral considerations in doing, the deliberative quest for the wisdom of the chosen actions. Phronesis, for Aristotle, was the knowledge linked to practicing morality.

Following Gadamer (1989), action is linked to practical wisdom and, as phronesis, always interlaced with the application of understanding. Application provides understanding with a direction. It defines the moral, in a specific case, in relation to collective understandings of right and wrong. Application is always present as an open opportunity for seeing things differently. Through the embodied aspects of morality (Merleau-Ponty, 1962), we build on a different rationale of phronesis (Lindberg & Olofsson, 2005), or practical wisdom, than seems to be present within the reflective practitioner paradigm (Clandinin, 2002; Clandinin & Connelly, 1999; Noel, 1999). For us, understanding is always focused by its application, and by values that frame moral actions, like teaching, and not only by critical reflection and reason.

Transferring this understanding of phronesis to the distance-based teacher training programme in question, the work of the teacher trainers becomes more than, or perhaps different from, ensuring that the teacher trainees develop a theoretical understanding of teaching and learning. If teacher training becomes a question of developing the skill of making moral decisions in unpredictable situations within the classroom, and since this particular teacher training is provided on a distance basis, questions arise concerning fostering and sharing notions of societal values. What are the possibilities for a communication technology to become an educational technology through which teacher trainees can develop these skills?

Education and Technology

The use of technology for educational purposes has undergone major changes during the last 50 years. From being primarily concerned with transmission or delivery, the changes have passed a cognitive focus on representation, and later on construction, and have gone towards a focus on social theories of learning and collaboration (Koschmann, 1996). In particular, the Internet has changed the way technology is used for educational purposes (Palloff & Pratt, 2003; Stephenson, 2001).

The idea of being part of an online education marks a shift from earlier views of the learning process. Focus has shifted from outcomes, in terms of students' performances as a product, and more to the view that today emphasises democratic learning, often in terms of participation, engagement, motivation, and ownership (Imel, 2001; Ó Murchú & Sorensen, 2003) and presence (Anderson, Rourke, Garrison, & Archer, 2001; Garrison & Anderson, 2003).

Distance-based tertiary education has also experienced this shift (Keegan, 1996). It has gone from being an educational form used to overcoming structural constraints such as time and place, to being an educational form characterised by two-way communication among a network of learners (Garrison, 2000). Distance education is no longer equivalent to correspondence education. Rather, it should be seen as an interactive learning experience supported by the use of ICT (Vrasidas & Glass, 2002).

Students that have had difficulties attending educational programmes are newly given the opportunity to participate (Brown, 2001). Teacher training is no exception to this trend. Students are recruited from all walks of life, making them more heterogeneous than previously. Accordingly, new values and beliefs are constantly being added to the melting pot of teacher training. A different practice is in the making whenever the traditional universities, with their emphasis on theoretical knowledge, offer professional programmes at a distance. One issue immediately raised is how do such courses address the normative, fostering, and value-based elements of education? How do teacher trainees learn that teaching has a moral aspect, as well as being a technical activity? These issues underlie the pedagogical rationale of teacher training in Sweden.

Teacher Training in Sweden

Teacher training has been a vocational programme since 1842. However, it was not until 1977 that it became a part of the university system. In the same year, distance education also became an integrated part of the university system as universities reached out to sections of Swedish society that had been denied university access. Although the possibility to study via distance-based teacher training has been available for nearly 20 years, most teacher training has been on campus and not online.

The 1999 proposal for teacher training reform in Sweden stressed the importance of teachers developing an understanding of the norms and values of the Swedish society as a means of establishing their personal ethic as a schoolteacher. Thus, they need to appreciate the value/framework based, for example, on the democracy, freedom, multiculturalism, and equity that is central to Swedish life.

Likewise, the parallel proposals on distance education identify the value of ICT-based flexible learning unconstrained by place or time. Insofar as such forms of distance education are adopted in teacher training, it must be assumed that the norms and values of Swedish society should necessarily occupy a central position. It can be assumed that the same values that permeate Swedish society should also permeate Swedish teacher training.

Education and Community

Sergiovanni (1999) claims that communities arise from shared values, sentiments, and beliefs. Humans are bonded together in an oneness, a set of shared values and ideas. In this sense, both universities and schools can be considered social organisations. It aims to create a we, rather than an I culture. Humans are joined together by being part of webs of meaning that tie humans together. They are no longer fulfilled by their individuality, but rather, by their commonality.

According to Bauman (2001), the word "community" has a double meaning. On the one hand, it gives an image of security and belonging; on the other hand, it offers a sense of place. Wenger (1998) suggested that communities share a history, and develop a shared repertoire of practices through the negotiation and fostering of such meanings. Palloff and Pratt (2003), like Wenger, suggest that community entails developing personal identity and shared values.

The concept of community is often used in connection with online learning. The community is considered to be located on the Internet, and is frequently described in terms of an online learning community (Carlén & Jobring, 2005; Haythornthwaite, 2002; Olofsson & Lindberg, 2005; Seufert, Lechner, & Stanoevska, 2002), or a virtual learning community (VLC) (Daniel, Schwier, & McCalla, 2003; Lewis & Allan, 2005; Schwier, 2002). Carlén and Jobring (2005) describe three main types of online learning community: educational, professional, and interest related. In this chapter, we focus on an educational online learning community. To qualify, certain characteristics are required, such as a curriculum, or other kind of steering document, related to its activities. Further, it can be assumed that students involved in those activities have the intention to learn (Nolan & Weiss, 2002). Additionally, within an educational OLC, students demonstrate their commitment to community by their joint participation, collaboration, and dialogue.

Participation, Collaboration, and Dialogue in an Educational OLC

In shaping the practice of online learning, teachers and students participate primarily through written text (Lock, 2002), organised around computer-mediated communication (CMC) (Garrison & Anderson, 2003). Different Web-based conference systems build online learning environments, and participation is possible both asynchronously and synchronously (Chong, 1998; Schwier & Balbar, 2002).

The definition of participation often excludes passive participation conducted by, for example, lurkers. Those are only reading or eavesdropping and not taking an active part in the ongoing dialogue. Many guidelines, design principles, or theoretical elaborations attempt to find methods that actively contribute to reducing the possibility of lurking or eavesdropping (see Sorensen & Takle, 2004).

Two potential strategies for dealing with the question of lurkers are, for example, ropes courses (Lowell & Persichitte, 2000) and initial bonding (Haythornthwaite, Kazmer, Robins, & Shoemaker, 2000). Ropes courses and initial bonding provide the members with opportunities to come together at the start of the education, and to build up a kind of social capital (Schwier, Campell, & Kenny, 2004) that holds the community together. This creates feelings of belonging, safety, and trust, and works as a foundation for future participation and activity within the educational OLC (Haythornthwaite, 2002; Hossain & Wigand, 2004).

This definition of participation, in terms of exclusion of lurkers, is closely aligned with a view of learning as an active process, where different aspects of knowledge building (Sorensen & Takle, 2002) are in focus, and where both active participation and close collaboration is considered to lead to learning (Dennen, 2000; Ingram & Hathorn, 2004; Mitchell & Sackney, 2001) and meaning making (Stahl, 2003).

To be able to both participate and collaborate in an educational OLC, with the aim of negotiating meaning and fostering a shared understanding among the members, some kind of asynchronic or synchronic dialogue is often used. This dialogue becomes a tool for learning, construction of knowledge, and creation of understanding (see Muukkonen, Hakkarainen, & Lakkala, 2004). This means that understanding built on dialogue provides the member in an educational OLC with an opportunity to contribute and be part of a process in which learning becomes a joint and collaborative venture.

Understanding through dialogue aligns us again with Gadamer (1989). By using dialogue, the possibilities for understanding are open for further questions, instead of answering questions with reproductive and predetermined content. Gadamer's approach brings up other issues concerning, for example, examination and normative features of education. It opens for a transformation of the university's more traditional role as producer and custodian of knowledge (Grundy, 1999). It also enables for an understanding of an educational OLC in which joint participation, as being, is central (Olofsson & Lindberg, 2006).

In Heidegger (1962), we find emphases on the unavoidable participation and presence in the world, which is always a shared world, a kind of being-together. Being together enables a common ground of existence, it enables a joint participation. This view of participation thereby includes the kind of passive participation conducted by, for example, lurkers within an educational OLC, and participation with other humans is considered a part of life as important as breathing.

In the sense of community, humans share history, and being is constructed and created as identities with certain and specific meanings in relation to shared repertoires negotiated and fostered in practice (Wenger, 1998). To participate with others raises questions about being together with others in a moral sense, and how to negotiate shared values. In the following, we will try to elaborate upon these questions by using the concept of phronesis in relation to data from an ICT-supported distance-based teacher-training programme.

Collecting and Analysing Data: Constructing the Pedagogies of Today

The empirical part of this chapter is based on an analysis of interviews with university teachers, all active teacher trainers within the ICT-supported distance teacher-training programme in question.

Programme and Participants

The programme was three-and-a-half to four-and-a-half years long, depending on the level of exam (which ranged from kindergarten teacher to upper secondary school teacher). The programme was distance-based, with compulsory gatherings two to four times each term at the university campus, or at the local learning centre nearest the teacher trainees' homes. These gatherings ranged from two consecutive days up to a week. Between gatherings, teacher trainees worked in study groups together, with a university teacher acting as a tutor.

The work between gatherings was conducted using a Web-based conference-system, WebCT, including both asynchronic and synchronic functions such as e-mail, chats, notice boards, and so forth. This Web-based conference-system formed a Web-based platform to which the teacher trainees had continuous access. The courses in the programme had their own web pages within the platform, and the study groups created their own educational OLC within the platform. The study group was intended to provide the teacher trainees with the social context for their studies between gatherings, and the Web-based platform was intended to provide the teacher trainees with an online learning environment.

The teacher trainers continuously encouraged the teacher trainees to actively use the Web-based conference system. This was, for example, stressed in the study guides handed over to the teacher trainees. Further, the tasks, designed by the teacher trainers, often required the teacher trainees to use ICT (chats, e-mail, and so on) in order to be solved (for example, comment upon each others' electronic portfolio and participation in online seminars).

In the programme, several activities were built around an idea of a social dimension permeating the programme, and that the Web-based conference-system or ICT should provide for the creation of a social dimension within this distance-based teacher training programme. In other words, the pedagogical rationale, or model, behind the programme included an idea of supporting a collaborative and social learning process using ICT.

The teacher trainers interviewed were chosen to represent as many university departments and courses as possible within the programme. The courses given were in the areas of social science, natural science, and the humanities. In this particular programme, the teacher trainees met one or more of the teacher trainers for at least half of the programmes' duration. Of interest might be that several of the teacher trainers were more or less unfamiliar with how to organise teacher training at a distance with the support of ICT. Since this was known in advance, it was addressed with courses and mentoring in distance education techniques for teacher trainers, as a way to make sure that the teacher trainers were well prepared prior to the start of the programme.

Interviews and Analysis

In the interviews, focus was on two major themes. Main interest was of issues concerning what the teacher trainers regarded as important aspects of teacher training and distance education, and issues regarding how they worked with the teacher trainees. The teacher trainers were given the interview guide in advance, and they had time to reflect over the questions before the interview (one teacher trainer, however, received the questions at the time of the interview). The interviews lasted for about 30 to 45 minutes, and were recorded. The interviews were thereafter transcribed, and given back to the teacher trainers for comments and correction. The commented transcripts are the data used for analyses.

In the analysis, we have built upon the concepts of affordances and constraints in line with Hutchby (2001), where the empirical use, and the possibilities for developing an understanding of how technological artefacts become involved in everyday conduct, are central. In other words, the technology affords certain forms of interactions and constraints others.

The descriptions below are abstractions based on the data collected. The descriptions should not be seen as an objective account of what occurs or does not occur within the programme: they are constructions of the pedagogies the teacher trainers talk about and use in the programme on different occasions and in different courses. They are also accounts of the aspects the teacher trainers mention as important for distance-based teacher training, but appear to carry out without the support of technology. In short, technologically afforded pedagogies include the talk about what teacher trainers do using technology, and technologically constrained pedagogies include the talk about what teacher trainers do without the use of technology.

Pedagogies that Appear to be Afforded by Technology

The following is a description of the constructed pedagogies used by the teacher trainers when applying technology to ensure active participation, required collaboration, and necessary dialogue.

Enabling Participation

Participation in the program is structured through the use of technology, for example, using different ways of distributing information. Including accessible study guides in the Web-based conference system, providing calendars with timetables for the courses, and by thinking about issues of information and access, the teacher trainers try to design for a supportive structure:

...when I design Web-based courses I always include a lot of contact points into the course, so that the teacher trainees have to send in an email or a memo or have been part of a discussion group... (Teacher trainer, humanities course)

The structure provided by technology also seems useful to the teacher trainers in providing clarifying information, and it helps them when responding to questions from single teacher trainees, as well as giving answers to the group as a whole. It could be used to provide direct answers in certain Web-based conferences, or to possibly build an FAQ (frequently asked questions) of teacher trainee difficulties:

...it was a good conference-system and the best thing was that you could reach everyone very easy. We could add questions from every person... (Teacher trainer, social science course)

The teacher trainers also emphasise the quality of the text-based information. The teacher trainers seem to see the need for clear and informative study guides, and a need to think about differences in text-based communication, so as to provide some kind of internal structure for the communication:

...because the writing is so special it is rather difficult to handle actually, and sometimes I even check, could I write this? Partly, so that the content

is factual but also if it feels like a friendly answer. For that might matter I assume... (Teacher trainer, social science course)

The structure given in the Web-based conferences is well thought through. For several teacher trainers, the intention of the programme is perceived to provide for groups of people living in remote and sparsely populated areas with varying home conditions, and to whom tertiary education has traditionally been a distant alternative. The teacher trainers have these aspects in mind when designing support for individual teacher trainees:

...at the same time, you have these people, with a lot in their baggage, and at the same time as there is some kind of alienation towards academic studies ... and it is important I think how you adapt or design the content to fit these categories of teacher trainees... (Teacher trainer, social science course)

The use of technology provides a structure for the programme, enabling these teacher trainees the possibility to attain a university degree and a vocational education. Although the opportunity to reach the teacher trainees as a group over a considerable geographical distance is considered as an advantage of using technology, it is sometimes not so easily achieved, since it is also associated with a lot of planning and coordination:

...we were supposed to have lectures and I was trying to arrange, coordinate schedules, it was terrible the way we had to carry on... (Teacher trainer, social science course)

At the same time, the emphasis on participation and dialogue also raises questions. There are several aspects of the teacher trainers' role that differs within a technology-supported distance-based programme. Perhaps aspects of the professionalism of the teacher trainers themselves come into focus:

...and then, I know a couple of teachers who think like, when it comes to teaching at campus they prepare very little, and not that they give bad lectures, but they kind of keep their knowledge in their heads, they can stand there and lecture, but how do you do that at a distance... (Teacher trainer, humanities course)

This opens up for an emphasis on the teacher trainer as facilitator and mentor of learning processes, and for a structure of the work within the programme to be designed in line with a student-centred approach focused on collaboration.

Working with Collaboration

Between gatherings at the university, teacher trainers use technology to enable collaboration in study groups. This is done, for instance, by providing the teacher trainees with Web-based conferences of their own:

...we create rooms for the teacher trainees where they can talk, it is important right, although it is rather a different matter whether they work or not, but the possibility is there, and in some groups, there can suddenly be added like hundreds of little comments... (Teacher trainer, natural science course)

The work carried out in the study-groups is adapted to a group work mode, and is expected to enhance learning at an individual level. The general feeling seems to be that group work is beneficial to all, but that it requires a different focus: a focus on the discussions and not on the products of the work:

...it is not the same course as a campus course, in that you have to think about the study-groups, to create tasks and assignments that fit the discussion forum... (Teacher trainer, humanities course)

This collaborative working mode is, therefore, not always chosen for assessment by the teacher trainers. Since the working mode is based on discussions, and there is less focus on product, teacher trainers exclude it from the ordinary forms of examination. However, the group work mode leads to questions concerning the form and role of assessment and examination:

...if it is as I have thought that the learning aspect of a certain examination form is the most dominant or important then you could ease up on the control aspect and it gives a lot in return, I give for example different kinds of home assignments and group work, and there too that you have to trust the teacher trainees that they do this because they are interested... (Teacher trainer, natural science course)

The earlier stated clarity that is needed in study guides and information seems not to be as apparent in the group work. The support and guidelines on group work seems to be less emphasised. Only some of the teacher trainers mention strategies for working in groups as being part of the overall structure, for instance, by establishing contracts within the study group to regulate the individual efforts, by circulating roles within the study group, or being a role model for discussions. What seems to be a general opinion is that dialogue has to be established.

Establishing Dialogue

The teacher trainers see a need to establish a dialogue with, and among the teacher trainees. An issue several teacher trainers mention but express mixed, if not sceptical, opinions about is the use of technology for giving lectures. Opinions voiced claim both technical and social difficulties in relation to lecturing:

...that is [give lectures] at the same time simultaneously in another [local city] so that we could link by videoconference and it didn't work... (Teacher trainer, social science course)

Technical difficulties are not the only problem, since other aspects connected to lecturing through videoconferencing are also mentioned. For example, the troublesome nature of the videoconference makes it more appropriate for monologues and as an information channel, rather than for dialogues among the participants:

...I could say from what I have seen of the Web-based that this with videoconferences is in many cases poor lecturing, and then I don't think we use the advantages of the computer, I mean we could just as easily have sent out a videotape since there is no dialogue it is still a monologue and we might as well have handed it out in writing... (Teacher trainer, natural science course)

Teacher trainers also use technology to establish dialogue with individual teacher trainees concerning content matters. In the courses, the teacher trainees frequently hand in assignments and use e-mail to ask questions about the theoretical content in the courses. Using this practice, teacher trainers both provide content matters as such, and didactical aspects of the content matter taught:

...with this email communication, you get opportunities in a totally different way or I don't know who is given the opportunity, the teacher trainees above all I assume, to show what they cannot and thereby get help... (Teacher trainer, humanities course)

Sometimes, in discussions concerned with aspects of the content, the discussions end up with issues or aspects that the teacher trainers had not themselves thought of:

...we have tried different models with chat models and some other stuff and it is obvious that sometimes you see the discussion taking turns you would never have anticipated... (Teacher trainer, natural science course)

Within the Web-based conference, dialogue is ongoing in the discussion forums, and the teacher trainees have access to these on a study group basis. However, in relation to teacher training as a governmental concern, with aims that must be reached and assessed, teacher trainers sometimes feel themselves forced into situations where they see themselves as having the responsibility to initiate and guide both the discussions and the content in chats, forums, or in seminars. The dialogue between teacher trainees and the dialogue with the teacher trainers is considered very important for reaching desired goals, but is, at the same time, paradoxical since, as mentioned above, the discussion might take unexpected turns:

...sometimes we join one of these discussion-groups we have created in the conference-system and try to steer them towards the goals... (Teacher trainer, natural science course)

In these ways, teacher trainers enable the teacher trainees to participate, using collaboration among the teacher trainees as a means, and ensuring that a dialogue is possible between teacher trainer and teacher trainee and between teacher trainees themselves. Thereby, the teacher trainees are regarded as being connected to an educational OLC, centred in the practice of teacher training. It seems, though, to be a practice primarily directed at the individual, and conducted mainly using different kinds of text-based communications. In short, the pedagogies afforded by the use of technology are focused on the theoretical perspective of teacher training; the content matter.

What are the Afforded Pedagogies?

The afforded pedagogies seem to be more in-line with the use of technology in the first three paradigms of Koschmann (1996). Teacher trainees receive information through the technology, and then they elaborate and reflect on the content, and produce signs of these elaborations and reflections in text. The educational OLC seems to be problematic when it comes to assessing qualities and aspects that are not content matter. We find this to be the case in the theoretical aspects of OLC literature, for example when passivity and lurking are unwanted forms of participation. If active participation is desired, in-line with a constructivist assumption about learning (or perhaps mainly since activity is more easily assessable when it comes to learning), then the passive aspects of learning a practice, of moving from the peripheral participation (Wenger, 1998) towards the centre of the practice, is overlooked.

Further, collaboration is seldom the basis for assessment, but is a mean of learning. This restricts the possibility for using the collaborative features and dialogue for educational intentions. A technologically afforded pedagogy, though not always articulated in a positive manner, involves the use of text. Text is often assessed from an individual perspective as a cognitively produced and reified thought (Anderson, Greeno, Reder, & Simon, 2000). If social aspects of collaboration are to have a chance of making their way into the teacher training programmes in practice, then this division ought to be considered. In the next section on pedagogies constrained, we elaborate further on this issue.

Pedagogies that Appear to be Constrained by Technology

The following is a description of the pedagogies used by the teacher trainers during the gatherings at the university, but for which they do not use technology.

Enabling Socialisation

When participating at the gatherings, teacher trainers' intentions are that the teacher trainees develop social skills and interpersonal capacities, as well as a readiness and capacity to handle complex and unpredictable situations in the classroom:

...they learn very much at the gatherings, and perhaps not primarily by sitting for eight hours in class at lectures, but by the total interaction with

the others in the group, which they cannot have when they are at home... (Teacher trainer, natural science course)

The socialisation that is present in group participation seems difficult for the teacher trainers to re-create without the gatherings, likewise, certain aspects of assuming a teacher's role. It seems that teacher trainers view this process as something that the teacher trainees have to experience in a face-to-face situation that the teacher training programme has to provide insights into:

...I'm not totally convinced, that you could replace the physical meetings altogether with ICT, I think that, and above all in a vocational program where you actually educate people to work with socialisation actually is in need of its physical meetings which would mean meeting between people in both an informal as well as an formal meaning, that this is actually socialising for the people who are going into a profession... (Teacher trainer, social science course)

It seems that teacher trainers regard learning as a social enterprise that is important for future teachers, and that discussion in study groups, physical meetings, and seminars are vehicles of learning a view of life:

...all people are part of all, in larger situations these situations have roots back and it is always a question of a cultural situation in some way and this, is about really, we humans can have a deeper dimension of being understanding of how things in the world are really connected... (Teacher trainer, humanities course)

This is an understanding in which values become part of everyday life and practice, and as such, they become part of the teacher-training programme, and thereby under the teacher trainers' control. The teacher trainers express an apparent need for careful consideration before deciding if these aspects are to be incorporated into the teacher training without being part of the gatherings at the university.

Working with Values and Beliefs

When considering the aims and goals of participating at the gatherings, the teacher trainers express several values, beliefs, and views that the teacher trainees should embrace ; a view of themselves as competent, active, and self-reliant learners. This implies a view of individuals as responsible for their own situation and actions, and it seems to the teacher trainers very much to be a question of forming the teacher trainees as persons:

...we also want to be clear in both the study-guide and when we introduce courses of our view of learning and our view of knowledge and our view of humans and so, right, and in that I also feel that we are saying, it usually amounts to a bit of a discussion and questions of what it means and so on that the responsibility lies very much with the learner... (Teacher trainer, social science course)

This is apparently an ongoing process that is not only concerned with the work conducted by the teacher trainees when they are studying at a distance, but also with the work carried out during the gatherings. The ongoing fostering of the teacher trainee as independent and self-reliant is also built into the work at the gatherings:

...the problem may also concern the gatherings having to be a bit different to strengthen the issue of the independent studies, to strengthen this community so that there is a parallel learning process rather than just being about the content we try to pass on... (Teacher trainer, social science courses)

Establishing Teacher Trainees as Teachers

Teacher trainers see their role as that of catalysts for challenging assumptions about both learning and subject matters that influence the teacher trainees at an individual level. This could be a way of looking beyond the 'here and now" of the programme towards the teacher trainees' futures as teachers, or as a way of questioning the experiences teacher trainees might have from periods of student teaching and reflecting on their chosen actions. It could raise the possibility of developing a deeper understanding of the content matter, and a view of potential difficulties future pupils may have, as well as the assumptions of the teacher trainees themselves:

...then connecting to the experiences met during periods of student teaching, having the time to stop and listen, how did you do then, why is that do you think... (Teacher trainer, social science course)

When working as a teacher trainer, there follows a responsibility to educate the teacher trainees to be good persons and good teachers. The teacher trainers might feel, for example, that the teacher training as an institution has this responsibility, or they might find themselves deciding who is to be given enough help to manage their studies. There even seems to be an underlying assumption that not all teacher trainees are suitable for a job as a teacher:

...well, it is a vocation with responsibilities to be a teacher and then I feel that it is important questions to keep in mind who we educate on an individual level too. It is a responsibility that we at the university have, that at the same time as I say yes that this will be very good pedagogues, well others you might feel less certain about and then you have to think about how in which way they should come prepared and then you look more to the individual level... (Teacher trainer, social science course)

These are issues the teacher trainers see a need for. Collaboratively, with the teacher trainees, creating a teacher-training programme that is connected to the expected practice of teaching, where the teacher trainers can assume the responsibilities of the university to educate the teacher trainees with both a capacity to teach as a skill, as well as fostering values aligned with those of the national curriculum.

What are the Pedagogies Constrained?

In the description of the pedagogies that are included at the gatherings, we find aspects of teacher training that are based on working with humans. It seems as if fostering values is possible only during the gatherings at the university. The forms of working seem to be more collaborative, and more in-line with a student-centred approach based on social theories of learning. Sharing the characteristics of the forth paradigm of Koschmann (1996), the one he sees as emerging based on collaboration and social theories of learning, these pedagogies seem to contain more of those aspects of becoming a teacher that are in-line with the knowledge of phronesis.

The concept of phronesis is concerned with being good, being able to choose and to deliberate upon different means and ends in relation to life as a whole. Using phronesis as the knowledge base that is needed to handle the complex and unpredictable teaching situation, and thereby also linking actions to moral deliberations, seems to be difficult for the teacher trainers using technology. Above all, it seems to be a question of creating a way for the teacher trainers to be present with the teacher trainees when they have experiences that ought to be challenged, and to be able to point out the assumptions they might have that should be questioned. This is both in relation to the teacher trainees' own assumptions, and in relation to their future pupils' assumptions and conceptions of the world.

Teacher trainers also seem to need to have the possibility to identify those teacher trainees who ought not to become teachers, and they feel a need to do so by challenging and questioning the teacher trainees in person. They also see a need for the teacher trainees to become socialised into the profession, to be able to take on the teachers' role and act as teachers in the future classroom. Above all, socialising teacher trainees to work themselves with socialisation requires the teacher trainees acting as their own role models. It seems that the teacher trainers have difficulties in realising these aspects of the teacher training in ways other than using the gatherings at the university.

We believe that the teacher trainees, by being part of a teacher training programme, are moving from being novices in the community of teachers, that is, being only peripherally involved, towards becoming full members (Lave & Wenger, 1991) and being initialised in the discourse of the practice (Wenger, 1998). The teacher trainees are learning to talk as professionals, and also learning to embrace the values that underlie the practice of teaching, but they do this as it is constructed within teacher training.

It seems as though the fourth paradigm of Koschmann (1996) is more present in the pedagogies that are technologically constrained. Working with humans is based on assumptions and values, and being part of a community is to share those beliefs. Even though the teacher trainers seem to include few aspects of working with the values common to Swedish society through the use of technology, the teacher trainees are still becoming members of a community. In the next section, we give some possible solutions and recommendations on how to include these aspects, by connecting the recommendations to the knowledge in action of phronesis.

Beyond the Pedagogies of Today

We believe that the work carried out at the gatherings could also take place between the gatherings, but that it involves using a pedagogy afforded not only by the text-based conference systems, but also by other kinds of technology than are used today. This means moving teacher training towards being a more student-centred rather than teacher-centred practice, and a belief in the use of more collaborative approaches. We can, by posing the question whether teacher trainers and teacher trainees necessarily have to meet physically if there is to be a fostering of common values, formulate the following afforded uses of technology, where meetings can be arranged through the Internet.

- Firstly, we suggest that teacher trainers and teacher trainees have access to the teacher trainees' experiences when teaching in classrooms, through the use of Web-cams and the common use of software for video conferencing.

The idea is to situate the Web-cams within different classrooms, and broadcasting live through a streaming server on to which the community members are logged into. This could open new possibilities for moving beyond the "talk" towards the "walk." It could create possibilities for a collaborative dialogue around different values that should embrace the future work for the teacher trainees. This could further create an opportunity to see how a future college deals with the different situations that emerges in classrooms, and also provides an opportunity for discussing the solutions from different aspects or perspectives. This opens the door to the future classroom and the fostering of phronesis, the knowledge of a practived moral.

- Secondly, we suggest that the Web-cams could be used to conduct "live role-play." To set up and act out different dilemmas in a classroom, related to the teaching profession, and to use educated teachers and pupils as "actors."

The teacher trainees could watch the play online, and thereafter discuss, either through a videoconference or a text-based chat, the actions taken by, for example, the teacher. This creates a dialogue between the teacher trainees and the "actors" (the teacher and the teacher trainee). It would also be possible to include the teacher trainers in the discussion. Phronesis is always interlaced with the application of understanding, and by conducting a reflective and critical dialogue around different scenarios played out, and by the actions taken by the "actors," it could provide an understanding of the morality involved in teaching.

- Thirdly, we suggest that digital cameras could be used by the teacher trainees to take snapshots of the practice they experience, and use them for reflection and discussion.

Organising snapshots from different schools and different contexts, and publishing them within the conference system used in the teacher training, can provide a sense of understanding for the diversities involved in teaching. Furthermore, it could display the differences of conditions framing the teaching, in terms of learning environments and pupils, and make this the common ground for a negotiated meaning among the members of the educational OLC. The teacher trainees could reflect differently depending on the meanings invested in the current situation displayed on the snapshot. In the dialogue, the teacher trainees have the chance to collaboratively negotiate a shared meaning, built on phronesis, of what the snapshot is all about.

- Fourthly, we suggest that digital video cameras could be used by teacher trainees to document certain ethical, as well as teaching, dilemmas, and later discuss them with their teacher trainers or with their study-group.

If certain ethical, as well as teaching, dilemmas are organised thematically, and highlighted in relation to aspects of the content matter in the university courses, different aspects of the knowledge base behind teaching could be brought to life within the collection of video clips available on the Internet. These might include issues regarding gender in relation to classroom discourse in social science, or related to discourse in natural science. Gender issues could also generate questions about other ways of defining social justice, and place those questions in a context of real life. Such questions that touch upon the application of understanding, the embodied aspects of morality, foster a readiness to work as a teacher.

- Fifthly, we suggest that digital video clips could make up a directory of good teaching experiences, as well as a directory of dilemmas associated with different subject areas.

If the video clips are organised on the basis of the content matter, it could help teacher trainees to identify whether the pupils have difficulties in certain areas. Collaboratively in the study-group and with the teacher trainers, teacher trainees can design learning strategies and teaching approaches towards difficulties that are (near to) authentic. This way of using technology allows an interlacing of theory and practice, and to work towards best practices of those solutions. This method of working opens for moral considerations in teaching, with trajectories towards the future work with pupils with special needs, or just those simply in need of support.

- Sixthly, we suggest that virtual reality (VR) would enable situating the teacher trainees in digital classrooms, and the provision of (near to) authentic teaching experiences.

The teacher trainees are provided with complex, real life dilemmas by using VR. For example, the teacher trainees could be placed in different classroom situations where different conditions are inherent due to the pupils' social background, the actions taken by the teacher trainees, the facilities in the classroom, and so on. The teacher trainees could be given the task of giving a lecture for the pupils, and the teacher trainers could change the circumstances of what happens in the classroom from one time to another. This provides a kind of social presence, and in a way, forces the teacher trainees to use knowledge in action. The technology thereby contributes to the fostering of phronesis. This integration of theory and practice is conducted in a safe and trusted learning environment, where the teacher trainees also have the possibility to always rethink their actions taken, and over and over again, step into the VR-created classroom to face new pupils and new real life dilemmas.

Ethical Considerations

Establishing these kinds of pedagogies causes a number of ethical problems. It is, for instance, difficult to obtain a licence to publish real-life situations and stories on the Internet, especially when there are children involved. The issues of integrity and ethical aspects of distributing individuals and their learning difficulties worldwide on the Internet is a troublesome aspect. But on the other hand, this could also incorporate the possibilities of involving the pupils themselves in the reflections and deliberations of teacher trainers and teacher trainees in a more active and democratic way. It could be a way to avoid student teaching periods becoming a practice *on* children, but rather a practice *with* children. We do not intend to provide solutions to these problems in this chapter (since we believe that there are no easy solutions), but one point we still want to make is that the issue is better handled if it is part of a democratic process of dialogue, and not silenced.

Any Problems Left?

Could these recommendations provide teacher trainers with possibilities to foster aspects of teaching aligned with both knowledge in action such as phronesis, and reaching the aims of teacher training associated with the values common to Swedish society? What remains to be a legitimate question for further thought is whether the practice of teacher training is different from the practice of teaching itself? In other words, are there any aspects of teacher training that involve learning to talk the talk, to which there is a different walk than teaching? If so, what are the legitimate grounds for teacher training? To avoid yet other questions, in the final part of this chapter we intend to propose another way to conceptualise teacher training through the use of technology that is more radical, and perhaps involves a more thorough change.

Future Trends: Bridging Communities

In a future perspective, technology affords teaching in tertiary education to include aspects that are common in vocational programmes. Vocations are basically practices, and as such, practices are fostering their participants into, not only a discourse of a professional (a talk), but an embodied knowledge, and an approach to the vocation of a professional (a walk). In teacher training, much of the focus in the national steering documents is concerned with ensuring that future teachers are aware of the values common to Swedish society. Teacher training is a moral arena, and as such, a state-controlled moral formation of the prospective teachers. Here, technology can afford the teacher trainers to develop practices where this formation could take place. The values in focus here are, for example, democracy and freedom, multiculturalism, and equity.

We can envisage a development of the educational OLC to include within the practice of teaching, more professionals with a relation to the teacher profession. Our idea is to widen the perspectives, and to put forth a more complex picture of the values that will embrace the teacher trainees' future work. Connected through asynchronic, as well as synchronic participation, collaboration, and dialogue, we can envisage the development of a shared teacher identity, made possible, if we expand the scope of the community from teacher training to teacher practice, by including already practicing teachers, teacher trainers, and teacher trainees in an (for all parts equally) educational OLC. Using Web-cams and software for videoconferencing, this bridged community could be realised, if sustained membership could be attained over time as a trajectory of participation into the practice of teaching.

With the help of technology, then, there is a real chance that the walks of life might affect the talks of academy. That the trajectory of participation of a teacher, the life of a professional among professionals, starts at teacher training, includes teacher training and the academic society and the workplaces in a true community of practice (Wenger, 1998), sustained and upheld in the virtual teaching of online learning, and thereby dissolving the notion of separate communities.

Conclusion

One main issue that hereby is addressed, is "Which community teacher trainees belong to?" Is it the university community, the teaching professionals' community or a community of teacher trainees? Who has the power to decide which belongings should have priority, and which communities should count? Since belonging to a community of a shared practice is the foundation to learning in a meaningful way, the question arises "Where is meaningfulness created"? Is it academically meaningful or future professional meaningfulness? Is it the practice of teacher training, that is being able to complete university studies, or is it the practice of teaching? Is the practice of teacher training the same as the practice of teaching? These questions all have one answer in a bridged community of teaching.

Our belief is that the use of Web-cams, software for videoconferencing, and the possibility to connect members of an educational OLC to each other, are technologies that could enhance the learning not just for the teacher trainees. It would enhance the practicing teachers' own teaching practices, and work as an in-service teacher training by including aspects and interests of the same community. Being a teacher in Sweden, in a future perspective, then becomes an aspect of a shared history of participation and a shared negotiated meaning. To make this happen, technology, and how it is used in teacher training, is more crucial than ever.

References

Anderson, J. R., Greeno, J. G., Reder, L. M., & Simon, H. A. (2000). Perspectives on learning, thinking and activity. *Educational Researcher, 29*(4), 11-13.

Anderson, T., Rourke, L., Garrison, D.R., & Archer, W. (2001). Assessing teaching presence in a computer conferencing context. *Journal of Asynchronic Learning Networks, 5*(2), 1-17.

Aristotle (1980). *The Niceomachean ethics.* (Sir David Ross, Trans.). Oxford: Oxford University Press. (Original work published 1925)

Bauman, Z. (2001). *Community: Seeking safety in an insecure world.* Cambridge: Polity Press.

Brown, R. E. (2001). The process of community-building in distance learning classes. *Journal of Asynchronic Learning Networks, 5*(2), 18-35.

Carlén, U., & Jobring, O. (2005). The rationale of online learning communities. *International Journal of Web Based Communities, 1*(3), 272-295. Retrieved May 13, 2005, from http://www.inderscience.com/search/index.php?action=record&rec_id=6927&prevQuery=&ps=10&m=or.

Chong, S. M. (1998). Models of asynchronic computer conferencing for collaborative learning in large college classes. In C. J. Bonk & K. S. King (Eds.), *Electronic collaborators: Learner-centered technologies for literacy, apprenticeship, and discourse* (pp. 157-182). Mahwah, NJ: Lawrence Erlbaum Associates.

Clandinin, D. J. (2002). Storied lives on storied landscapes. Ten years later. *Curriculum and Teaching Dialogue, 4*(1), 1-4.

Clandinin, D. J., & Connelly, F. M. (1999). *Shaping a professional identity: Stories of educational practice.* New York: Teachers College Press.

Daniel, B., Schwier, R. A., & McCalla, G. (2003). Social capital in virtual learning communities and distributed communities of practice. *Canadian Journal of Learning and Technology, 29*(3), 113-139.

Dennen, V. P. (2000). Task structuring for online problem based learning: A case study. *Educational Technology & Society, 3*(3), 329-336.

Gadamer H. G. (1989). *Truth and method.* London: Sheed and Ward.

Garrison, D. R. (2000). Theoretical challenges for distance education in the 21st century: A shift from structural to transactional issues. *International Review of Research in Open and Distance Learning, 1*(1), 1-17.

Garrison, D. R., & Anderson, T. (2003). *E-learning of the twenty-first century. A framework for research and practice.* London: Routledge Falmer.

Grundy, S. (1999). Partners in learning. School-based and university-based communities of learning. In J. Retallick, B. Cocklin, & K. Coombe (Eds.), *Learning Communities in Education* (pp. 44-59). London: Routledge.

Haythornthwaite, C. (2002). Building social networks via computer networks: Creating and sustaining distributed learning communities. In K. A. Renninger & W. Shumar (Eds.), *Building virtual communities—Learning and change in cyberspace* (pp. 159-190). Cambridge: Cambridge University Press.

Haythornthwaite, C., Kazmer, M. M., Shoemaker, S., & Robins, J. (2000, September). Community development among distance learners: Temporal and technological dimensions. *Journal of Computer-Mediated Communication, 6*(1). Retrieved January 25, 2005, from http://www.ascusc.org/jcmc/vol6/issue1/haythornthwaite.html.

Heidegger, M. (1962). *Being and time.* Oxford: Blackwell.

Hossain, L., & Wigand, R. T. (2004, November). ICT enabled virtual collaboration through trust. *Journal of Computer-Mediated Communication, 10*(1). Retrieved January 25, 2005, from http://jcmc.indiana.edu/vol10/issue1/hossain_wigand.html.

Hutchby, I. (2001). *Conversation and technology. From telephone to the internet.* Cambridge: Polity Press.

Imel, S. (2001). Learning technologies in adult education. *Myths and realities. ERIC Clearinghouse on Adult, Career, and Vocational Education,* (17). Retrieved February 8, 2005, from http://www.cete.org/acve/docs/mr00032.pdf.

Ingram, A. L., & Hathorn, L. G. (2004). Methods for analyzing collaboration in online communications. In T.S. Roberts (Ed.), *Online collaborative learning: Theory and practice* (pp. 215-241). London: Information Science Publishing.

Keegan, D. (1996). *Foundations of distance education.* London: Routledge.

Koschmann, T. D. (1996). Paradigm shifts and instructional technology. In T. D. Koschmann (Ed.), *CSCL: Theory and practice of an emerging paradigm* (pp. 1-23). Mahwah, NJ: Lawrence Erlbaum Associates.

Lave, J., & Wenger, E. (1991). *Situated learning: Legitimate peripheral participation.* Cambridge: Cambridge University Press.

Lewis, D., & Allan, B. (2005). *Virtual learning communities. A guide for practitioners.* Berkshire: Open University Press.

Lindberg, J. O., & Olofsson, A. D. (2005). Phronesis: On teachers' knowing in practice. Towards teaching as embodied moral. *Journal of Research in Teacher Education, 12*(3), 148-162.

Lock, J. V. (2002). Laying the groundwork for the development of learning communities within online courses. *The Quarterly Review of Distance Education, 3*(4), 395-408.

Lowell, N. O., & Persichitte, K. A. (2000, October). A virtual course: Creating online community. *Journal of Asynchronic Learning Networks, 4*(1). Retrieved January 25, 2005, from http://www.aln.org/publications/magazine/v4n1/lowell.asp.

Merleau-Ponty, M. (1962). *Phenomenology of perception.* (C. Smith, Trans.). London: Routledge & Kegan Paul.

Mitchell, C., & Sackney, L. (2001, February 24). Building capacity for a learning community. *Canadian Journal of Educational Administration and Policy, 19*. Retrieved February 1, 2005, from http://www.umanitoba.ca /publications/cjeap/articles /mitchellandsackney.html.

Muukkonen, H., Hakkarainen, K., & Lakkala, M. (2003). Computer-mediated progressive inquiry in higher education. In T. S. Roberts (Ed.), *Online collaborative learning: Theory and practice* (pp. 28-53). London: Information Science Publishing.

Noel, J. (1999). On the varieties of phronesis. *Educational Philosophy and Theory, 31*(3), 273-289.

Nolan, D. J., & Weiss, J. (2002). Learning in cyberspace: An educational view of virtual community. In K. A. Renninger & W. Shumar (Eds.), *Building virtual communities: Learning and change in cyberspace* (pp. 293-320). Cambridge: Cambridge University Press.

Olofsson, A. D., & Lindberg, J. O. (2005). Assumptions about participating in teacher education through the use of ICT. *Campus Wide Information Systems, 22*(3), 154-161.

Olofsson, A. D., & Lindberg, J. O. (2006). "Whatever happened to the social dimension?" Aspects of learning in a distance-based teacher education programme. *Education and Information Technologies, 11,* 7-20.

Ó Murchú, D., & Korsgaard Sorensen, E. (2003, September 23-26). "Mastering" communities of practice across cultures and national borders. In *Proceedings of the 10th Cambridge International Conference on Open and Distance Learning*, Cambridge, UK. In A. Gaskell, & A. Tait (Eds.), *Collected Conference Papers*. The Open University in the East of England Cintra House. Retrieved February 23, 2005, from http://www2.open.ac.uk/r06/conference/Papers.pdf.

Palloff, R. M., & Pratt, K. (2003). *The virtual student: A profile and guide to working with online learners*. San Francisco: Jossey-Bass.

Schwier, Richard A. (2002, June 1). *Shaping the metaphor of community in online learning environments*. Paper presented at the International Symposium on Educational Conferencing. The Banff Centre, Banff, Alberta. Retrieved January 9, 2005, from http://cde.athabascau.ca/ISEC2002/papers/schwier.pdf.

Schwier, R. A., & Balbar, S. (2002). The interplay of content and community in synchronous and asynchronous communication: Virtual communication in graduate seminar. *Canadian Journal of Learning and Technology, 28*(2), 21-30.

Schwier, R. A., Campbell, K., & Kenny, R. (2004). Instructional designers' observations about identity, communities of practice and change agency. *Australasian Journal of Educational Technology, 20*(1), 69-100.

Sergiovanni, T. (1999). The story of community. In J. Retallick, B. Cocklin, & K. Coombe (Eds.), *Learning communities in education* (pp. 9-25). London: Routledge.

Seufert, S., Lechner, U., & Stanoevska, K. (2002). A reference model for online learning communities. *International Journal on E-Learning, 1*(1), 43-55.

Sorensen, E. K., & Takle, E. S. (2002). Collaborative knowledge building in Web-based learning: Assessing the quality of dialogue. *International Journal of E-learning, 1*(1), 28-32.

Sorensen, E. K., & Takle, E. S. (2004). A cross-cultural cadence. Knowledge building with networked communities across disciplines and cultures. In A. Brown & N. Davis (Eds.), *World yearbook of education 2004. Digital technology, communities & education* (pp. 251-263). London: Routledge Falmer.

Stahl, G. (2003). *Meaning and interpretation in collaboration*. Paper presented at Computer Support for Collaborative Learning (CSCL 2003), Bergen, Norway. Retrieved January 18, 2005 from http://www.cis.drexel.edu/faculty/gerry/cscl/papers/ch20.pdf .

Stephenson, J. (Ed.). (2001). *Teaching & learning online. Pedagogies for new technologies.* London: Kogan Page.

Vrasidas, C., & Glass, V. S (Eds.). (2002). *Current perspectives on applied information technologies: Distance education and distributed learning.* Greenwich, CT: Information Age Publishing.

Wenger, E. (1998). *Communities of practice. Learning, meaning and identity.* Cambridge: Cambridge University Press.

Chapter III

Enhancing the Design of a Successful Networked Course Collaboration:
An Outsider Perspective

Rema Nilakanta, Iowa State University, USA

Laura Zurita, Aalborg University, Denmark

Olatz López Fernández, University of Barcelona, Spain

Elsebeth Korsgaard Sorensen, Aalborg University, Denmark

Eugene S. Takle, Iowa State University, USA

Abstract

This chapter presents a preliminary critique of an online transatlantic collaboration designed for collaborative learning. The critique by external reviewers using qualitative methods within the interpretivist paradigm hints at critical factors necessary for successful online collaborative learning. The evaluation seems to support the view that in order to raise the quality of online dialogue and enhance deep learning, it is good practice to heed, as well as give voice to participants' needs by involving them directly in the design of the course. This has the potential to enhance student motivation and learning. The authors plan to continue their work, and present a more grounded and detailed evaluation in the near future involving multiple data sources, comprehensive surveys, and document analysis.

Introduction

Online learning, or e-learning, is witnessing unprecedented growth in higher education today. The term "online learning" implies learning facilitated mainly through network technologies. Garrison and Anderson (2003) describe the growth of e-learning as explosive and disruptive. However, the authors contend that empirical research on the potential and actual use of e-learning technologies seems to be lacking. "To date, published research and guides consist of innumerable case studies and personal descriptions and prescriptions but little in the way of rigorous, research-based constructs that lead to an in-depth understanding of e-learning in higher education" (Garrison & Anderson, 2003, xi).

As a response to Garrison and Anderson's call for rigorous research on e-learning, we direct our attention, in this chapter, to the pedagogic design of online collaborative learning. Using empirical data, we critique an active, online, collaborative-learning initiative that involves two distinct courses. Instructors of the respective courses have published their work in academic journals as a case of innovative online collaboration (Sorensen & Takle, 2002, 2004 a, b; Takle, Sorensen, & Herzman, 2003). The pedagogic design is characterized by principles of collaborative knowledge-building (Scardamalia & Bereiter, 1996) and communities of practice (Wenger, 1998). We analyze the collaboration and offer recommendations. It is important to note that the evaluation and recommendations presented herein constitute the preliminary stage of evaluation. Data involved in our analysis is limited to few sources, namely, an unstructured end-of-course call for student comments, one interview, and a few publications. We intend to continue this work of gathering data from multiple sources, and plan to present a detailed and more grounded evaluation in the near future.

In spite of the nascent state of our critique, we believe our findings and recommendations have implications for pedagogic design of online collaborative learning initiatives. Our findings, albeit preliminary in nature, stress the need to revisit current perspectives and representations of collaborative learning on the Web. This chapter, thus, questions the effectiveness of current e-learning models that claim to promote collaborative learning, and challenges us to explore alternative strategies and models.

The chapter is divided into five sections: (1) a brief literature review on the importance and value of collaborative learning; (2) description of an online collaboration, undertaken by two instructors on either side of the Atlantic, with expertise in two separate disciplines. (This section presents the instructors' story outlining the evolution of their respective courses and the relationship between them.); (3) the evaluators' story, their involvement in the project, their preliminary analysis, tentative recommendations, and foreseeable challenges in implementation; (4) future course of action; and (5) conclusion.

Literature Review

Collaborative learning, according to Pea (1994), is a process that can be perceived as transmissive/transactional or transformative. In the transactional viewpoint, the goal of collaborative learning is "one of creating good strategies in pedagogic practice to make sure the right information is transmitted or in neo-Piagetian terms, *constructed* by the learner…[while in the transformative perspective] not only students but also teachers are transformed as learners by means of their communicative activities" (Pea, 1994, p. 289). Pea maintains that the transactional model of learning has dominated educational practice and advises, "this one-way view of information conveyed by authorities to knowledge produced in (not by) the learner must be enriched for a vital education to occur" (Pea, 1994, p. 288).

Dillenbourg (1999) takes a different approach, and maintains that collaborative learning has different definitions. However, he argues that it can be understood "along three dimensions: the scale of the collaborative situation (group size and time span), what is referred to as 'learning,' and what is referred to as 'collaboration'" (Dillenbourg, 1999, p. 2). Collaborative learning can, therefore, be understood narrowly as people working together on joint projects solving problems and generating new knowledge as a result of it. Or it can be understood broadly as an ongoing process of acculturation by which individuals become members of groups and communities by participating in community activities, and acquiring skills and knowledge characteristic and valuable to the group. The latter perspective echoes Lave and Wenger's (1991) understanding of learning in general. The authors contend that learning is fundamentally social, and entails a gradual induction of the individual into his/her community of practice through interacting with community members.

However, Dillenbourg also stresses the activity of individuals working together, alone, does not characterize collaborative learning. On the contrary, he believes that collaborative learning works because of the *activities* individuals perform when they collaborate, and these activities trigger learning mechanisms. Dillenbourg therefore prefers to define collaborative learning as "a *situation* in which particular forms of interaction among people are expected to occur, which would trigger learning mechanisms, but there is no guarantee that the expected interactions will actually occur" (Dillenbourg, 1999, p. 5).

Salomon, alternatively, focuses on the degree and depth of dependency between collaborating participants as central to collaborative learning. He believes true collaborative learning is characterized by genuine collaboration, which is a condition of "'genuine interdependence' between individuals that calls for sharing knowledge/information, adopting complimentary roles, a 'pooling together of minds'" (Salomon, 1992, p. 64).

In spite of differing perspectives on collaborative learning, scholars agree that collaborative learning has the potential for developing critical thinking skills, helping students develop a social support system, and promoting good citizenship. These arguments have helped migrate to the Web collaborative-learning practices popular in classrooms such as joint problem solving facilitated by debate and discussions. This has led to a proliferation of collaborative virtual environments such as Wikis and Web logs, groupware, and discussion forums.

With this background information in mind, we now turn to the description of an ongoing online collaboration whose design essentially reflects the core principles of collaborative learning discussed previously. The online collaboration has been termed successful based on student feedback, and the fact that it has been in existence for almost a decade, and continues to flourish with enthusiastic support from students, faculty, and the administration.

The Collaboration So Far:
A Description of Two Courses

Since 1999, two courses, and their designers, have been collaborating, each in their own way, to promote and practice learning in virtual environments. The two courses include a Danish distributed CSCL (computer-supported collaborative learning) course (from the humanities) on how to design teaching and learning in pedagogically appropriate ways using ICT-technology, and an American mixed-mode CSCL course (on-campus and Web-delivered) (from the sciences) on global environmental issues (Sorensen & Takle, 2004a, 2004b).

The Danish Course: Master Education in ICT and Learning (MIL)

The MIL course was one of three courses (and one project work) in a one year distributed CSCL university education program (within the humanities) for high school teachers, and for people from the educational system of organizations. The one-year education program was offered as continuing education on a half time basis. The goals of the course were for students to be able to integrate ICT appropriately in teaching and processes of development, and, at a high level, to be able to guide and implement the use of ICT in teaching and learning, as well as in other organizational contexts.

The American Course: Global Change (GC)

The GC course was a conventional course (within the sciences) for senior undergraduates or beginning graduate students at a U.S. university. It was migrated to the Web in 1995, with new features being added, as ancillary software became available. The instructors also introduced learner-centered activities in place of, or supplemental to, conventional lectures. The goals of the course are (1) to help students come to an understanding of the interconnectedness of the global environment, and the role of humans in charting (by design or by default) its future trajectory; (2) to instill an appreciation for, and recognition of, authoritative literature on global-change issues; (3) to engage students, within the course and across national and cultural boundaries, in dialogue on global-change issues, including ethical issues; and (4) to develop skills, individually and collaboratively, for implementing the knowledge-building process as an approach to problem solving.

Nature of Collaboration

The Danish course was designed on Wenger's principles of social learning (Wenger, 1998). The American course, especially the design of the discussion forum, gradually evolved, guided by the same framework. According to Wenger, learning is fundamentally social, and is supported through communities of practice. Learning is enabled when individuals are gradually initiated into a community of practice (CoP) and engage in its activities, generating artifacts that subsequently impact their learning (Lave & Wenger, 1991; Wenger, 1998). The three concepts central to a communities of practice framework, namely, mutual engagement of participants, a joint enterprise, and a common repertoire (understandings, conventions, values, tacit skills) were taken into consideration as the course designs evolved over the years. With each iteration, changes were made to the GC and MIL courses. The changes addressed instructor observations and student feedback, as well as CoP principles. We will now briefly describe the design of both the courses, as informed by Wenger's CoP principles.

Joint Enterprise

The collaboration between MIL and GC courses was characterized by joint enterprise, namely the enhancement of the GC course from a directed pedagogy to one supporting collaborative knowledge-building. The collaboration was designed, fundamentally, as an exploratory learning experience benefiting both parties: The GC design team needed sound and qualitative expertise input to enhance the design of their Web-based course, and the MIL students held to gain "virtual practice" and authentic, meaningful learning through investing their professional expertise critiquing the GC course. Moreover, in relation to ICT, the MIL course further aimed to explore two learning technologies, the Web and videoconferencing, and their potential for enabling such collaboration and functioning as tools for collaborative global dialogue (Scardamalia & Bereiter, 1996): a potential to ensure global collaboration towards supporting the development of a democratic world.

The focus on joint enterprise was also intended to lead to a true transcendence of both geographical and conceptual borders. First, it implied a transcendence of geographical boundaries, enabling knowledge construction, dissemination, and access to learning resources across the Atlantic. Second, the collaboration crossed the strong and traditional borders between two disciplines: the sciences and the humanities. Finally, the collaboration crossed the strong and traditional borders between American and Danish pedagogic traditions of course design and delivery. American tradition, in the context of this study, alludes to direct instruction stressing transfer of information from teacher to student. Whereas the Danish tradition indicates a culture of "democratic participation through dialogue," assuming, to some degree, that learning typically emerges when students/participants engage in debate and dialogue.

Mutual Engagement

The collaboration allowed for active participation by both MIL and GC students. GC students were encouraged to comment on course design and the design of the virtual portfolio, and offer constructive suggestions for improvement. For MIL students, the course offered a dimension of authenticity at different levels.

Firstly, the MIL students explored, via their learning assignments, the features of the GC course (i.e., simulations, virtual portfolios, multiple choice quizzes, etc.), through their own individual GC portfolios, and collaborated with each other within the GC context. In addition, on the behest of the GC designer, they also provided advice, as pedagogical experts on the design features of the GC course. They especially focused on the enhancement of (1) the pedagogical techniques and tools used in the GC course, and (2) the features of the virtual portfolio for supporting global dialogue and collaboration in learning.

1.   **Pedagogical techniques and tools in GC course.** The instructors designed an exercise for the Danish students, working within the context of their course on ICT and pedagogical methods, to work also within the context of the Global Change course as a basis for evaluating its functionality and pedagogical methods. These evaluations were done with the help of technologies used in both courses: virtual portfolios from the GC course, and videoconference from the Danish course. Each student was issued a password-protected electronic portfolio as a launching point for exploring three features of the course, namely the use of quizzes and class summaries for encouraging integrative thinking, use of simulations as a means of allowing open-ended hypothesis testing, and use of the electronic portfolio as a personal space for managing interaction with the course. MIL students used their portfolios to post their evaluations through both private comments to the instructors, and through public postings, by which they engaged in dialogue with other MIL students and both the instructors.

2.   **GC virtual portfolio features.** The collaboration between MIL and GC courses implied a kind of "virtual practice" for the MIL students. The Web-based platform for the GC course was developed completely in-house without use of commercial software, and the developer was available and responding to inquiries 7 days per week. This allowed MIL students to not only suggest software changes, but to see them implemented within 24 hours, provided such changes did not impede the ongoing flow of the course. The course, therefore, served as a hands-on laboratory for instructional designers, allowing for exploring software implementations of pedagogical alternatives without themselves being software experts.

A Shared Repertoire

We (course designers/instructors) had hoped that working collaboratively with each other and the students would generate shared understandings of teaching and learning, tangible artifacts such as papers and documents on online course design, an improved functional online portfolio system, and a shared understanding of the respective content areas.

Problem: Lack of Depth in Dialogue and Collaboration

Overall, it is fair to say that over the 6 years, the course has turned out to be a success. By success, we mean that the students, especially the MIL students, have expressed a high degree of satisfaction with the course. We attribute this, on the one hand, to the innovative ideas behind the design of the collaboration, and on the other hand, to the deep engagement of the collaborating professors, technical designers, and MIL students, as it evolved each year.

However, the same is not the case with the GC course. Despite structural changes in how online dialogue was managed, and incentives provided, rigorous tests of GC student learning revealed that the GC course fell short of its goal of becoming a model for true collaborative learning (Sorensen & Takle, 2002; Takle et al., 2003). "The resulting discussion could most accurately be described as superficial remarks spiced with personal agendas, religious fervor, cutesy comments, and an occasional spark of academic interest" (Takle et al., 2003). The requirement to demonstrate critical-thinking skills by identifying the nature of each post only resulted in a "collection of monologs that lacked the linkages required by our definition of collaborative learning" (Takle et al., 2003). Although GC student evaluations document satisfaction with the course as a whole, their feedback still reflected dissatisfaction on the use of techniques such as dialogue and collaboration.

As help toward addressing this problem, we employed three PhD students to help us throw new light on, and produce new ideas as to how to address this issue. All three students work with exploring and evaluating new models of online learning including virtual portfolios, as a central element in their research dissertations. In this chapter, we (course designers/instructors and the evaluators) wish to jointly explore, analyze, and discuss, from the perspective of social learning theory, potential causes, and solutions for this. The next section describes the critique undertaken by the evaluators.

A Critique: An Outsider Perspective

The three PhD students who critiqued the MIL course come from different institutions and contexts, but share a common interest in collaborative design and evaluation of online environments. Two of them are especially interested in understanding the role of electronic portfolios in teaching and learning and are concerned with ways to enhance its current practice, keeping in mind issues of design, assessment, and equity. This section is divided into four parts: (1) an introduction to the evaluators, (2) a brief description of methods used for analysis, (3) a preliminary analysis, followed by (4) a brief discussion of the findings and tentative recommendations for improvement along with challenges for implementing the recommendations.

The Evaluators

The evaluators represent PhD students who became acquainted with each other through an international collaboration between universities in the U.S. and Europe. This collaboration was funded jointly by an EC-US grant. Two evaluators come from European universities (Spain and Denmark), and the third evaluator is enrolled at the same Midwestern U.S. university where the GC course is offered.

The evaluators possess overlapping research interests. The evaluator from the U.S. is interested in the design and development of pedagogically sound online learning environments. Her main interests lie in the area of instructional design. She is interested in applying social theories of learning, specifically, the theoretical framework of communities of practice, to the design of online learning environments. She regards the electronic portfolio's dynamic spaces as possessing high potential for supporting learning that promotes high-order thinking, professionalism, and citizenship.

The evaluator from Spain is interested in the use of digital portfolios for learning and assessment in higher education. Her main interest lies in understanding how this innovative alternative method of assessment is pedagogically impacting the agents involved in the educational process, especially the learners at the university. She has implemented a descriptive study to conceptualize and define pedagogical criteria for assessing digital portfolios in higher education. She has obtained interesting results in relationship with its potential for social learning, with the objective of promoting learner autonomy.

The evaluator from Denmark is interested in user involvement and empowerment in education, specifically in online education. She is working with methods of introducing the user's voice into the design and evaluation process of online environments.

Methodology

We (the evaluators) argue for the need to use qualitative research methods for analysis. We take an interpretivist approach to understand the particular situation and/or the context of the problem. In the interpretivist paradigm, reality is always socially constructed (Willis, Jost, & Nilakanta, in press). Hence, understanding the context "is not a single understanding of the 'right' way of viewing a particular situation. It is, instead, an understanding of multiple perspectives on the topic" (Willis et al., in press). In our case, it therefore becomes important to understand the participants'-students' and instructors'/designers'-points of view of the course.

Researchers working within the interpretive paradigm typically use qualitative sources such as reflection, open-ended interviews, and document analysis. Similarly, our data pool included interviews with the instructors, published articles on the collaboration, and student evaluations. Since we are at the initial stages of our analysis, the data pool remains small. We plan to cast a wider net in the future and collect interviews with students as well.

To be noted here is the fact that our analysis relies on data gathered mainly from the GC course. Since the problem, in our case, has been defined by the instructors as a lack of depth in GC student discussions, we limited our analysis, for the present, to GC student evaluations, interview with GC instructor, and articles published jointly by the two instructors.

Analysis and Findings

At the time of this report of preliminary results, we have yet to complete a detailed investigation. Nevertheless, an initial analysis was conducted with promising results. Using qualitative methods, we recognized common themes that echoed instructors' observations (Takle et al., 2003) regarding lack of depth in student dialogue in the GC course. By "lack of depth," the instructors imply a lack of negotiated meaning-making activity as illustrated previously.

A preliminary study of GC students' evaluations, and subsequent interview with the GC instructor shed some light on the reasons underlying this problem. It highlighted four issues that could probably explain the lack of depth in student dialogue, namely: (1) mismatched student expectations and course goals, (2) unfamiliarity with Web-based learning, (3) mandated number of online postings, and (4) mandatory quizzes.

1.   **Mismatched student expectations and course goals.** There was a general disconnect between GC student expectations of the course, and the course goals presented earlier under description of GC course. Students seem to expect a traditional, teacher-lead pedagogy, but were required to engage with other students in reflective dialogue and self-directed learning. The quote below is typical of student opinion of the course:

   ...I think the instructor should teach rather than having to learn from the comments of other students. (GC student evaluation, Spring 2001)

2.   **Unfamiliarity with Web-based learning.** The students were not used to an online format. This was particularly true of earlier iterations of the course. Below is a comment that represents a difference in learning styles, and probably uneasiness with Web-based learning:

   I hated the Web-based class. To help influence my critical thinking skills, I need to actually hear the information being presented to me. Dialogue does not accomplish in furthering my education. (GC student evaluation, Spring 2002)

   This particular observation needs to be further investigated. Personal interviews and focus group sessions with GC students should shed light on the above claims.

3.   **Mandated number of postings.** The requirement for a specific number of postings each week seemed to distract students from the authenticity of the task:

   I don't like the fact that we have to generate a certain amount of responces [sic]. It's hard to know what people will be responding to. (GC student evaluation, Spring 2004)

4.   **Mandatory quizzes.** The data indicated a distinct distaste for taking quizzes (tests). We speculate the distaste for quizzes, which students have to take frequently during the semester, may also be impacting their performance in, and motivation to, dialogue in depth:

Quizzes did not help me to learn, frustrating to complete multiple times each week. (GC student evaluation, Spring 2004)

The negative attitude toward quizzes seems to be the only consistent factor in all the evaluations. Despite the fact that most students value the quizzes as a means of encouragement for keeping up with course assignments, complaints against quizzes seem to overwhelm other complaints.

Discussion

The findings reveal philosophical differences between GC students and their instructor. We find this aspect the most challenging. It is well known that our beliefs impact our behaviour, and we consider a mismatch between GC student beliefs and course goals has the potential to lead to a lack of motivation, and an unwillingness to engage in reflective, collaborative thinking on the part of GC students. The findings appear to indicate that GC students expected a traditional pedagogic design, where knowledge is passed on from the teacher to the students. In contrast, the instructor, who designed the course, expected his students to take the initiative and engage in collaborative and transformational learning (Pea, 1994).

Our conversation with the GC instructor also indicated that in the earlier iteration of the GC course, students were mandated to learn collaboratively without being informed of its rationale and benefits. After subsequent student feedback expressing dissatisfaction with the course, the instructor informed students of the rationale for a collaborative approach, and subsequently, the number of complaints of this nature declined. However, our initial analysis does not indicate if this led to more substantive discussion or improved learning outcomes. This remains to be investigated.

We find students' dissatisfaction with pedagogic strategies, such as required number of postings and the use of quizzes, intriguing for two reasons. Firstly, these strategies undertaken by the instructor are directed, and represent traditional teaching strategies. They contradict instructor's motivation for a collaborative approach that requires students to take control of their learning. Secondly, by the same token, students should have liked the strategies, since they reflect their earlier expectations of a course. We believe further interviews with the instructor and students would help shed light on this contradiction.

Unfamiliarity with the tool, although easily addressed through preparatory training, can exacerbate challenges instructors face. It is quite possible that GC students' unfamiliarity with Web-based tools hindered them from exploring the virtual portfolio, resulting in an unwillingness to engage in rich dialog.

Although there is no empirical evidence as yet, we believe the findings allude to a lack of ownership of the course on the part of the GC students. The students consider themselves consumers of the course, demanding efficient and effective service. The resistance and distaste toward making a specific number of postings and taking quizzes demonstrates that students do not consider these requirements relevant and valuable. They view them as an undue obligation that has been imposed upon them externally. This stands contrary to the founding principles of the GC course, which calls for the development of community and shared ownership through active participation and collaborative knowledge-building. We plan to investigate this further in the next stage of our research.

Based on our initial findings, we now present our recommendations. The recommendations that follow consider the lack of ownership as a central issue that needs to be addressed first.

Recommendations

It is crucial to note here that the recommendations are primarily for redesigning the collaboration between the GC and the MIL course. However, most of the recommendations listed below pertain to the MIL course. According to the instructors, the MIL course possesses greater flexibility since it involves a smaller number of students, and therefore, will be easier to modify. Hence, we focus on a redesign of the MIL course as a way to enhance the online collaboration.

We also believe that, over the years, the two courses have become so tightly bound that improvements to the MIL course will eventually lead to changes in the GC course. Moreover, these recommendations are at best tentative, and should be taken as the first step to deeper reflections and suggestions on educative online practices.

We offer the following recommendations, informed by principles of CoP, namely, a common domain of interest/joint enterprise, mutual engagement, and shared repertoire:

1. **Joint evaluation of the course.** In order for GC students to "buy in" to the collaboration, we suggest MIL students evaluate the course in teams with GC students. Such collaborative work has the potential to draw GC students into a community of practice that includes MIL students as well as the course designers. We speculate, based on our preliminary analysis, the GC students do not see themselves as members of a community of practice engaged in a joint enterprise, namely, the improvement and enhancement of the GC course. They consider themselves as consumers and not owners

of the course, because they had no voice in its conceptual design. In spite of diligent efforts by the course instructor and technical support personnel to get student feedback, the students' voices remain muted since they provide "feedback" on a design whose parameters have already been set by the course designers. According to research on participatory systems design, this does not represent "true" participation (Ehn, 1992; Greenbaum & Kyng, 1995; Greenbaum & Madsen, 1993). In real participation, end users have substantial influence on the design, and hence, a feeling of ownership in the product. Moreover, heeding the end user's voice is not only democratically desirable, but also productive, as it results in better systems that are more user friendly, more adapted to the needs of the users, (De Young, 1996; Kuhn, 1996), helps facilitate early adoption by users, and is cost-effective in the long run (Hirschheim, 1985).

An evaluation of the course that involves both groups of students would help translate the findings into practice by breaking the dichotomy between subject and object. The end users (the GC and MIL students) come to represent both, the subject/agent and the object of the design. End users themselves help collect, process, and analyze the information using methods easily understood by them. Likewise, the knowledge generated is typically used to promote actions for change (Tilakaratna, 1990). We suggest, furthermore, that more course designers adopt this strategy. While much is being written about collaboration and empowerment of the learners, it does not seem to translate into involving students into the design process.

2.   **Gaining intercultural competence.** A collaborative evaluation will help expose both groups to a new discipline, a new culture and worldview. MIL students have expressed surprise at how much they learn about meteorology while evaluating the GC course. GC students similarly, with MIL students' help, will have to apply evaluation criteria that are pedagogically sound, thus exposing them to terms and concepts in the field of education. This will help nurture an interdisciplinary and intercultural orientation, which has been deemed crucial for a global society (Davis, Cho, & Hagenso, 2005).

Research shows that for a course to be successful, the teaching methods and the design have to be adapted to the content (Asenio, 2000). Presently, MIL students are expected to evaluate a course whose epistemology they are not familiar with, thus running the risk of judging it with a bias, without being aware of it. This also highlights a point necessary to heed while designing e-learning experiences. Users of technology are not, and cannot be, abstract humans isolated in a contextless environment. They are indeed embedded in their social and cultural contexts (Star & Ruhleder, 1996). It is not hard to imagine that the bigger the difference between the social and cultural context of the users (students and instructors) and the evaluators, the more difficult it is for the evaluators to evaluate which courses fit into the worldview, and the needs of the users (Whitmore, 1994).

3.   **Peer mentoring.** We suggest MIL students act as pedagogic advisors or mentors, guiding GC students in developing learner autonomy, ability to self-critique, and learning to collaborate with their GC peers to understand course content. Moreover, this collaboration has the potential to result in joint publications involving MIL and GC students.

4.   **Evaluation criteria.** We suggest MIL students evaluate the GC virtual portfolio based on an adaptation of the pedagogical criteria for digital portfolios (López, 2004). This will provide MIL students a theoretical framework to inform their evaluation, making it comprehensive and sound.

5.   **Discipline-based evaluation.** We suggest embedding the idea of the need to evaluate courses in their social, cultural, and disciplinary content in the MIL curriculum. Examples of different pedagogical strategies in different content areas across the humanities, social sciences, and natural sciences should be discussed as part of the MIL course content. A robust framework for course evaluation could then be developed.

6.   **Online pedagogic tool evaluation.** We suggest that when MIL students evaluate the GC virtual portfolio, or other pedagogic strategies in the GC course, they are also exposed to other examples of online pedagogic tools. This will help expand their knowledge and repertoire.

Challenge

The first four recommendations are crucial for enhancing GC dialogue. However, they call for collaboration between GC and MIL students. This is problematic, since the two courses are not held at the same time, and there is not much possibility of the situation changing, given the reality of two different educational systems and cultures. Therefore, we suggest that both parties look to developing course mentors drawn from pools of ex-students who have taken the respective courses and share a metaunderstanding of the collaborative process. These mentors will help maintain a sense of continuity, and help develop an identity for the community that includes students from both courses, and their respective instructors. In other words, these mentors will, in essence, function as "the elders" of the community, preserving community knowledge, and helping it evolve by initiating newcomers (the current students) into their community of practice.

There is, of course, the extra challenge of creating mechanisms that will ensure GC and MIL students' motivation to get involved in the extra work involved in this evaluation.

Future Course of Action

As noted in the *Introduction*, the evaluation presented in this article constitutes a preliminary study, but one that has already yielded important information that needs further verification. We plan to continue our work collecting more data from other sources such as interviews with current students and past students, academic advisors, and comprehensive surveys, in order to test our findings. We also look forward to exploring the recommendations in greater detail, to check for their feasibility, especially with regard to transforming the collaboration between the MIL and GC students.

Conclusion

With the proliferation of e-learning in higher education, it is important that we pay close attention to the design of online technologies and pedagogies that claim to support learning that is necessary for a global world: learning that aims to develop future leaders who are successful across cultures, disciplines, and geography. This requires not just a focus on the design of online courses, but also exploring innovative ways of evaluating them.

We presented one way of evaluating an online collaboration. The evaluation, based on an unstructured, end-of-course call for comments by students, showed that designing for collaborative learning does not necessarily lead to expected transformative learning (Pea, 1994). In order for that to happen, we need to adopt a participatory design approach. All primary stakeholders, especially the students and the teachers, need to have an effective voice in the design of the course from its very outset. We presented recommendations and suggested ways to involve students in designing and evaluating the course as a means of fostering collaborative learning. We also propose that design of courses should not be a onetime task, but a recurrent, interactive activity involving all stakeholders (including students). This, we realize, increases the task load for already busy course designers, but it has the potential to increase course quality.

Although this chapter presents a particular instance of online collaborative learning, its findings should be applied at the conceptual level. Willis et al. (in press) explain and advise:

Research adds to our understanding of different contexts and situations, but our application of that understanding is not a technical process, it is reflective. That is, we must thoughtfully make decisions in our own practice and those thoughtful decisions must be based on all our understanding. (Willis et al., in press, n.p.)

The findings of the critique presented in this chapter may concern one instance of improving an already successful online collaboration. However, it also helps shed light on the complexity inherent in collaborative learning initiatives. We also realize further research is required to verify our findings and test our assumptions. We plan do to this in the near future, and publish our findings for peer review.

Acknowledgments

The authors gratefully acknowledge the support provided by ILET (International Leadership for Educational Technology) in facilitating the collaboration that resulted in this chapter. ILET is an intercultural exchange program between six universities and four supporting partners from Europe and the U.S. It aims to build an international learning community, with a collection of expertise and resources to prepare teacher educators for the twenty-first century networked world. ILET is funded jointly by the European Union, and the Department of Education, U.S.

References

Asensio, M. (2000). Learning from failure: A case study of networked learning in management education. *Journal of the Institute of Training and Occupational Learning (JITOL), 1*(1), 77.

Davis, N. E., Cho, M. O., & Hagnson, L. (2005). Intercultural competence and the role of technology in teacher education. Editorial *Contemporary Issues in Technology and Teacher Education, 4(4).* Retrieved March 24, 2005, from http://www.citejournal.org/

De Young, L. (1996). Organizational support for software design. In T. Winograd, J. Bennett, L. De Young, & B. Hartfield (Eds.), *Bringing design to software.* Reading, MA: Addison-Wesley.

Dillenbourg, P. (1999). What do you mean by collaborative learning? In P. Dillenbourg (Ed.), *Collaborative-learning: Cognitive and computational approaches* (pp. 1-19). Oxford: Elsevier.

Ehn, P (1992): Scandinavian design: On participation and skill. In P. S. Adler & T. A. Winograd (Eds.), *Usability: Turning technologies into tools* (pp. 96-132). New York: Oxford University.

Garrison, D. R., & Anderson, T. (2003). *E-learning in the twenty-first century: A framework for research and practice.* New York: Routledge-Falmer.

Greenbaum, J., & Kyng, M. (1995). The design challenge: Creating a mosaic out of chaos. *Conference on Human Factors in Computing Systems. Conference companion on Human factors in computing systems,* Denver, CO (pp. 195-196).

Greenbaum, J., & Madsen, K. H. (1993). Small changes: Starting a participatory design process by giving participants a voice. In D. Schuler & A. Namioka (Eds.), *Participatory design: Perspectives on systems design* (pp. 289-298). Hillsdale, NJ: Lawrence Erlbaum Associates.

Hirschheim, R. (1985). User experience with and assessment of participative systems design. *MIS Quarterly, 9*(4), 295-304.

Kuhn, S. (1996). Design for people at work. In T. Winograd, J. Bennett, L. De Young, & B. Hartfield (Eds.), *Bringing design to software.* Reading, MA: Addison-Wesley.

Lave, J. & Wenger, E. (1991). *Situated learning: Legitimate peripheral participation.* Cambridge, UK: Cambridge University Press.

López Fernández, O. (2004, October 28). Evaluation criteria for digital portfolios: A pedagogical perspective for higher education. *II International Conference "ePortfolio 2004."* LaRochelle, France.

Pea, R. D. (1994). Seeing what we build together: Distributed multimedia learning environments for transformative communications. *The Journal of Learning Sciences, 3*(3), 285-299.

Salomon, G. (1992). What does the design of effective CSCL require and how do we study its effects? *ACM SIGCUE Outlook, 21*(3), 62-68.

Scardamalia, M., & Bereiter, C. (1996). Computer support for knowledge-building communities. In T. Koschmann (Ed.), *CSCL: Theory and prac-

tice of an emerging paradigm. Mahwah, NJ: Lawrence Erlbaum Associates.

Sorensen, E. K., & Takle, E. S. (2002). Collaborative knowledge building in Web-based learning: Assessing the quality of dialogue. *The International Journal on E-Learning (IJEL), 1*(1), 28-32.

Sorensen, E. K., & Takle, E. S. (2004a). Diagnosing quality of knowledge building dialogue in online learning communities of practice. *Proceedings of the World Conference on Educational Multimedia, Hypermedia and Telecommunications, 2004*(Vol. 2, pp. 2739-2745).

Sorensen, E. K., & Takle, E. S. (2004b). A cross-cultural cadence in e-knowledge building with networked communities across disciplines and cultures. In A. Brown & N. Davis (Eds.), *Digital technology communities and education* (pp. 251-263). London: RoutledgeFarmer.

Star, L. S., & Ruhleder, K. (1996). Steps towards an ecology of infrastructure: Design and access for large-scale collaborative systems. *Information Systems Research, 7*(1), 111-134.

Takle, E. S., Sorensen, E. K., & Herzmann, D. (2003, February 9-13). Online dialogue: What to do if you build it and they don't come? *Proceedings of the 12th Symposium on Education*. Long Beach, CA: American Meteorological Society.

Tilakaratna, S. (1990). *A short note on participatory research*. Paper presented at a seminar funded by the United Nations/ Government of Sri Lanka, Community Participation Programme 1988 to 1995 in Colombo, Sri Lanka, 6 January 1990. Retrieved March 23, 2005, from http://www.caledonia.org.uk/research.htm#Characteristics%20of%20 Participatory

Wenger, E. (1998). *Communities of practice: Learning, meaning, and identity*. Cambridge, UK: Cambridge University Press.

Whitmore, E (1994). To tell the truth: Working with oppressed groups in participatory approaches to inquiry. In P. Reason (Ed.), *Participation in human inquiry* (pp. 82-98). Thousand Oaks, CA: Sage Publications.

Willis, J., Jost, M., & Nilakanta, R. (in press). *Qualitative research methods for education*. Greenwich, CT: Information Age Publishing.

Chapter IV

Enhanced Learning and Leading in a Technology-Rich, 21st Century Global Learning Environment

Ian W. Gibson, Macquarie University, Sydney, Australia

Abstract

This chapter explores the evolution of thinking about learning, resulting from the increasingly ubiquitous presence of instructional technology and communications technology in learning environments. It provides a short history of the pedagogical growth of technology usage. It further describes the impact of technology on the potential transformation of four-walled classrooms into global, online learning communities from a constructivist perspective, while looking at learner/teacher roles in the learning process. The Global Forum on School Leadership (GFSL) and the Global Forum on Educational Research (GFER) are introduced as applications of interactive educational technology, suitable for twenty-first century learners, teachers, and school leaders, that emphasize creation of new knowledge using exploration and collaboration through self directed, technology-enhanced learning controlled by the learner. The intent of this discussion is to explore the impact of technology on learning, and recognize the transformative power behind introducing this learning experience into school leader preparation programs.

Introduction

This chapter explores the evolution of thinking about learning, resulting from the increasingly ubiquitous presence of instructional technology and communications technology in learning environments. It provides a short history of the pedagogical growth of technology usage. It further describes the impact of technology on the potential transformation of four-walled classrooms into global, online learning communities from a constructivist perspective, while looking at learner/teacher roles in the learning process. As examples of the type of collaborative learning that can result from this educational metamorphosis, the Global Forum on School Leadership and the Global Forum on Educational Research are introduced as Type II applications of interactive educational technology suitable for 21st century learners, teachers, and school leaders. Type II applications are those that emphasize creation of new knowledge using exploration, discovery, and collaboration, through the use of the computer as a self-directed learning tool controlled by the learner, rather than technology applications that parallel programmed-learning events, such as drill and practice exercises, which are considered to be Type I applications (Maddox, Johnson, & Willis, 1997).

The concept of the Forum Series brings together learners who share a common goal, a common subject area, or a common profession, and encourages them to interact and learn together. Among the many differences that learners bring to the learning task, the one central, and very obvious difference upon which the Forum Series depends, is culture. The Global Forum on School Leadership creates a learning partnership between a class of neophyte school leaders enrolled in a school leader preparation program in the United States, and a similar class of students enrolled in a school leader preparation program in Australia. The Global Forum on Educational Research is designed for a doctoral program, and replicates the success patterns of the GFSL, with the added variable of Web-based video interactivity. The discussion in this chapter begins with a strong focus on the evolution of technology use in schools, the concomitant application of pedagogy as the computer merged with classroom activities, and the related evolution of learning in increasingly technologically rich learning environments. The Forum Series is then introduced as a Type II application, capable of facilitating a self-directed, student lead orientation to learning, increasingly necessary in a learning world where a global focus is being forced upon learners, and where global understanding and awareness is becoming a basic survival skill. The chapter generates a global version of the five characteristics of Type II applications. On the way, it explores the necessity for a world-minded orientation to learning; delves into the idea of using networking, electronic learning, and communications technologies to break down communication barriers and concomitant power and control barriers; and concludes with an analysis of the impact of Global Forum type experiences on participants' own learning experiences and professional skills.

Understanding the Impact of Technology History and Learning History on Predominant Thought Patterns in 21st Century Education

In the late 1970s and early 1980s, some of those who were newly learning about the potential of the microcomputer, after its emergence from the garage, were intent upon using it to more efficiently do what had been done in classrooms for generations (Dwyer, 1996). In classrooms around the globe, in both developing and developed countries, students were learning multiplication facts, rote learning equations, repeating the content of the next chapter in the social studies textbook, struggling with spelling tests on Friday afternoon, and expectantly waiting for the teacher to tell them what they were to learn next, and how they were to think about the issues that were being presented to them. In these early years, the microcomputer was seen as the tool that would assist the teacher in accomplishing these tasks more efficiently. In retrospect, it was abundantly clear, during this period that technology applications of this type made "little use of the technology's potential interactive qualities and flexibility" (Sewell, 1990, p. 13).

Drill and practice, memorizing content knowledge, and rote learning were the "lingua franca" of this era of classroom activities. These same activities became the focus of early software developers, intent upon capitalizing on the educational potential of the microcomputer. With the advent of this new learning tool, the factory model of learning, which had emerged during the industrial era in the United States, was being incorporated into the early software developed for the microcomputer. Drill and practice, and page turning applications were effectively converting the computer into an efficient teaching tool that was replicating teacher-centered classroom activities prevalent in the large majority of classrooms. In the process though, the real power of the technology was being diverted away from its transformative potential, and away from deep learning processes. The didactic model of learning that these applications represented, temporarily usurped the real potential of this new learning technology.

For the first decades of the integration of the microcomputer into classroom activities, teachers who had learned their craft in an industrial age model were delighted that the computer was able to free them up from the "mind-killing" (O'Brien, 1994) tedium of drill and practice lessons. These teachers were praising the patience of a machine that could take even the slowest of learners and patiently reinforce them in the simple and superficial tasks of knowledge acquisition, and encourage them to rote learn at their own pace, without the shadow of an impatient teacher responsible for 30 other students, looming in the background. In this situation, technology was considered to be a "patient, nonthreatening tutor for basic skill acquisition ... offering students infinite opportunity to repeat problems until process or content [was] mastered" (Dwyer, 1996, p. 18).

During this period, the personal computer was considered by early adopting educators to be a remarkable educational tool for these reasons, and so, a bandwagon effect evolved from the early years of incorporating technology into school contexts. Educators, all over the country and the world, were convinced that this tool would be of benefit to learning. Not too many were aware of exactly how it would benefit learning, or even specifically why it was a good thing to spend education dollars on computers. Nor was much evidence gathered to confirm the impact of this new technology on the learning of students. Nevertheless, school boards and other decision makers were assailed with demonstrations of the latest drill and practice software, and encouraged to devote millions of dollars to the cause, often without anything other than a sense that some good would come from this new and powerful teaching tool. It was not long before some teachers and school leaders were convinced that the value of the tool did not depend solely upon the qualities of the tool itself, but on the learning objectives it supported (Maddox et al., 1997).

Return on the educational dollar investment on computers soon became a concern. Around this time in the process, the bourgeoning numbers of "technology coordinators" popping up around clusters of schools were able to boost the support for increasing expenditures on "learning technologies" by adding the idea that teachers and students who learned how to program these computers using BASIC, and later LOGO, were on the cutting edge of a process that would support the problem solving and logical thinking processes so valued within classroom walls. The bandwagon effect had come into its own as these arguments were adopted and propagated in school after school. Subsequently, teachers and school leaders around the globe boasted about the one standalone computer that now proudly stood in pride of place in school classrooms, and was able to be used by all students whenever the set tasks of the learning day had been completed. Other more advanced adopters of this technology gathered computers into labs, where a teacher directed an entire class of neophyte computer users (often sharing one computer between three or four users) in a technology-based learning process at a scheduled time, in a scheduled place, and with a specific, scheduled learning objective in mind. Computers were being adopted because they were replicating the traditional and familiar content and learning paradigm of the teacher-centered classroom.

Up to this stage, it had become clear that some educators had not devoted sufficient "careful thought to the kinds of teaching and learning tasks to which the microcomputer tool [could] best be applied" (Maddox et al., 1997, p. 18). Soon after the first "rush" of readily available personal computing power had mellowed somewhat, many thoughtful educators began to ask a very simple question: "What learning objectives are being supported by the use of a microcomputer in the classroom environment ... and is the investment worth it?" With a focus on achieving specified, measurable learning objectives as part of the question, technology coordinators began to develop answers using an orientation towards curriculum integration activities that looked like scope and sequence documents, indicating where computer technologies might best be integrated in order to serve the learning process. Software "application" programs began to appear as frequently as drill and practice programs in staffroom discussion and classroom activities.

Adventurous teachers began to think about using word processing tools to improve the writing habits of students. Even more adventurous teachers began to consider the hidden value of electronic spreadsheets in the manipulation of numerical data and the exploration of complex mathematical relationships, or database application software for allowing students to manipulate and reorganize the "facts" of history in order to generate new knowledge, new understandings, and a new perspective on the common understandings of history, economics, and society.

Some time later, when computers facilitated access to the world's data banks, and supported text, audio, and then video conversations between learners a room apart, and then half a world away from each other, the transformational potential of these information and communications technologies (ICTs) in learning environments began to burgeon. No longer was the personal computer relegated to a world of learning typified by drill and practice, stand-alone programming, and teacher-directed, full class activities. What began to emerge was a new paradigm of learning that focused upon the individual learner directing, managing, and being responsible for the products of his/her own learning, with the assistance and guidance of a coach who shared in the learning experience in a teaming, partnering capacity. In this learning context:

...investigations often lead to solutions, but always lead to discoveries. Student work [was] varied, oriented to sharing information, and critically reviewing ideas, models, and other peer products. Work completed here [was] open to public scrutiny, and this fact alone [was] sufficient to affect the quality of work done, the commitment to that work, and whether the work [was] conceived of as real or not by learners. (Gibson, 2001, p. 42)

Understanding the Shift from the Classroom Learning Environment to the Global Learning Environment

More recently, and decades after the advent of personal computing and its patchy integration into learning environments via the faltering steps of early adopting educators, information and communications technologies have been incorporated into a multitude of socioeducational activities and environments. Ubiquitous technology has provided a growing percentage of the population with access to more information than has ever been readily available in human history. This era is typified, again for the first time in human history, by a growing phenomenon, where new generations are teaching their predecessors the skills necessary to survive. In the process, the line, representing the traditional distribution of power and control in classroom environments between the teacher and the learner, is being redrawn. That once sacred and clearly defined sociocultural divide of learning, which had the teacher as the controller of learning, the provider, the instructor, the active participant in the teaching/learning process, was now not so clearly defined or distinct. The personal computer was exercising its potential for upsetting the balance of power in the classroom and in the learning process. The pendulum of power had begun to swing toward the learner, who now had the tools and the technological savvy to begin to construct his/her own meaning from experiences, explorations, and self-directed learning adventures, in collaboration with peers, with community partners, recognized experts, and with global colleagues. In once traditional classrooms, where the locus of power and control was clearly in the hands of the teacher, the computer had become a catalyst for breaking the power, authority, and control structures that had long predominated in the teaching and learning dyad, and was beginning to positively impact the way that learning was defined and practiced in technologically rich classroom environments. In fact, many were recognizing a gap between the way new generations were living and learning outside of school, and the way they were being taught and constrained to learn in schools (Partnership for Twenty-first Century Skills, 2002).

In recognizing these different approaches to learning, and the newer generation's reliance on information and communication technology as a means of making sense of their world, Breck (2004) has provided a treatise on the power of network science, and its impact on the distribution of knowledge and power in learning and political contexts. She described what she considered to be the answer to ending ignorance and separation in the learning world as a consequence of empowering all individuals on the planet with an interactive device, so that access to each other, to knowledge, and to the subsequent power derived from that access was evened out across the world's populations:

"Connectivity," Breck claimed, *"... is enormously threatening ... threatening in education because it centers communication in the student, not the teacher or administrator. It is threatening to all varieties of thought police because everyone can know everything.... Because it provides a global commons of human knowledge in which every person on Earth will share."* (Breck, 2004, p. viii)

She inferred through these comments that it is the responsibility of new generations of teachers to adopt a view of their roles to include these unique characteristics. Breck cajoles her readers, and teachers in particular, to adopt a global learning perspective, in order to create an environment where learners become globally aware, responsible, and empowered, and thereby able to display the world-mindedness necessary for the future in a new millennium where ubiquitous technology is responsible for the generation of incredibly rich and unconstrained access to each other, and the world's knowledge base, all at a moments notice. The native inhabitants of this new learning landscape, Breck claims, are rapacious consumers of real world, authentic, and meaningful interactions (with others and with knowledge construction). Those who teach them (the newcomers to this technology-rich environment), she says, remain reserved and hesitant observers of unfamiliar patterns of learning represented by multitasking, and new ways of doing.

Similar redistributions of power have been recognized in the global political arena, and have, likewise, been supported by the transformational potential of technology. Among many who have written about the empowerment of disenfranchised peoples around the world through technology, De Vaney, Gance, and Yan Ma (2000), have explored the true potential of information and communication technologies, in situations where the tyranny of power has been redefined through the nontraditional use of existing technologies, for the good of disenfranchised populations. In *Technology and resistance: Digital communications and new coalitions around the world,* these authors have explored the process whereby citizens without a voice have taken control of their own destinies, and through the creative use of communications technologies, have propagated a different type of revolution, by demanding that their voice be heard in the political spectrum of the times. The case studies contained in this book relate authentic situations where technology has created new personal potential, and increased the power available to the technology user in the global political arena.

In striving to understand the concerns of those represented by these case studies, another level of understanding, related to the global arena of influence and power, is uncovered. For many readers in the western world, these case studies represent a new reality, which, heretofore, has not forced itself into the western level of consciousness. For those readers in the United States who have lived under the choice of isolationism in previous generations of legislative policies, media coverage, and personal practices, a buffer zone has been created between those western realities, and the realities of the vast majority of the world's people. There has existed little reason for those under these policies to be cognizant of the world of differing attitudes outside that buffer zone. The Twin Towers in New York on September 11, 2001, Bali in October, 2002, Riyadh in the last months of 2003, Madrid in early 2004, and now the continuing tragedy of Iraq, and the spate of bombings in London and elsewhere in 2005, have permanently changed the appropriateness of that practiced complacency however, and confirmed that using the transformation potential of ICTs in a global learning environment is crucial to the creation of globally aware, responsible, and empowered learners capable of providing a positive direction for the evolution of the new century. Deeper applications of technology usage in the classroom, and on the world stage, have provided increasing evidence that deep thinking, deep creativity, and deep understanding must evolve from the transformational potential of information and communication technologies in personal and professional learning arenas. It is exactly these characteristics that Breck (2004), DeVaney et al. (2000), and others are extolling for the benefit of a new generation of teachers: a new breed of educators who are willing to risk being involved in the adventure of developing new skills and new practices for the improvement of a system of learning appropriate for the technology-rich environment of the new century.

In order for the revolution in learning capabilities to be fully incorporated into the system of the educational enterprise, however, it is clear that new generations of school teachers and school leaders must, themselves, experience alternative means of learning and thinking in order for them to develop the mental models (Senge, 2002) capable of capitalizing on the transformational potential of ICT use in learning environments suitable for the 21st century. If, indeed, the likelihood of the terrorist-based tragedies of the first six years of the new millennium can be ameliorated through greater understanding of alternative world views, and a deeper level of honor and respect being afforded those who think and act differently than we do (Breck, 2004; De Vaney et al., 2000; Rimmington, Gibson,

Gibson, & Alagic, 2004), then it is imperative that the Type II applications of technology discussed by Maddox et al. (1997), incorporate global awareness appropriate for the technologically rich learning context of the new century.

It is in this global and technologically rich learning context that the greatest good can be realized by:

- allowing learners the active intellectual involvement required as a starting point in the virtual exploration of the realities of others from around the globe;

- providing the learner control of the interactions that happen through the screen, and the focus of discussion beyond the screen, which are designed to engage learners in their own personal and professional development;

- encouraging the learner to communicate, share self-constructed files, text, voice, and video resources, and other inputs, as a commonplace and necessary feature of deep involvement in the individual and collaborative construction of meaning on a global scale and with a global audience;

- removing any limits to the creativity, approach, or intensity imposed on the learning processes or artifacts incorporated in exchanges; and

- recognizing that the resultant behavior of the learner and learning partners might, indeed, move beyond the standard hours of the school day in order to capitalize on the potential of new ICT-based learning opportunities through the exploration of global relationships.

Further, the full impact of this Type II experience might itself continue to develop and affect subsequent learning opportunities, and influence (change the direction of) personal growth and development of expertise and understanding beyond the initial experience (see Maddox et al., 1997, p. 27 for a listing of the original characteristics of Type II applications upon which this global reconstruction of the typology has been based).

In providing additional support for this contention, Ian Jukes (2000) suggested that integrating technology was far more than just using computers or allowing students to spend lots of time online. He claimed that effective integration occurred only when technology added real value to learning, and only when the resulting curriculum was based on the living concerns of those who studied it. In this setting, technology takes on the vastly different role of a tool, rather than a tutor. If used thoughtfully, this general-purpose tool is capable of providing learners with access to information, expert communications, opportunities for collaboration, and a medium for creative thought, expression, and knowledge construction (Gibson, 2001). ICTs provide the opportunity many have needed to share the responsibility of learning with their students, and create an unlimited and unrestrained student-centered learning environment.

Dwyer (1996) simplified the differences between teacher-centered (knowledge instruction) and student-centered (knowledge construction) learning environments through conceptualizing the role of technology in each of these settings. He found it useful to consider the computer as a tutor in the knowledge instruction setting, and as a tool in the knowledge construction setting (Gibson, 2001). In the majority of Type II applications, and by definition, the computer takes on the role of a tool that learners use to construct their own knowledge. However, to take school-based learning with technology beyond the realms of traditional, industrial age models of content acquisition, to student-centered engagement in authentic learning opportunities, a new vision of learning possibilities is necessary (Gibson, Schiller, Turk, & Patterson, 2003), and that vision must be shared, understood, and supported by the next generation of school leaders, if it is to become reality.

The Global Forum on School Leadership

An example of Type II applications of ICTs suitable for twenty-first century learners can be found in the Global Forum on School Leadership. As a Type II application of interactive computing technology, the Global Forum on School Leadership brings together learners who share a common goal, and creates a learning partnership between them. The "Forum" operates within a university graduate course context that distinguishes itself from traditional, administrator-preparation programs, and that focuses upon teamwork, collaboration, individual and collaborative construction of knowledge, the creation of professional communities of learning, integration of coursework, and heavy technology usage, as it applies to the redefinition of the roles of school leaders in schools designed to meet the needs of twenty-first century learners.

Dependent partly on the theoretical framework derived from problem-based learning (PBL) (Boud, 1985; Boud & Feletti, 1991; Bridges, 1992), this program contextualizes learning around authentic problems of practice that are explored in collaborative team settings, and lead to learner-directed and setting-enhanced learning. A large portion of program activities comprises "the research-based exploration of authentic, contextualized problems of practice in collaboration with administrators, teachers, and other educational personnel from local school districts" (Gibson, 2002a, p. 2). However, "contextualizing ...leadership experiences in traditional and unvarying cultural contexts often presents its own limitations, particularly during the formative period of leadership philosophies, perspectives, and practices" (Gibson, 2002a, p. 2). Subsequently, the focus on incorporating alternative leadership contexts into class discussions through a global orientation to leadership preparation, particularly in a post September 11, 2001 international context, became a necessary focus to course development and improvement.

Further, the program within which the Global Forum on School Leadership resides, and from which it draws its philosophical foundation, is dependent upon the belief that collaboration and peer-based learning reflects the realities of the professional workplace for which these students are being prepared. An investigation of the problem analysis process of PBL (De Grave, Boshuizen, & Schmidt, 1996) supports the notion that exposure to different ideas, in a team setting based on collaboration and shared growth, leads to conceptual change, and that "group interactions serve to encourage activation and elaboration of existing knowledge and integration of alternative views" (Gibson, 2002a). Moreover, this program epitomizes the growing trend recognized by Heath (1997) for models of education that accommodate a constructivist view of learning, incorporating the use of emerging technologies (Hannafin & Land, 1997). The simple logic here, as it applies to the Forum, is that interaction with peers and colleagues concerned with the same major professional issues, but having a different cultural slant to those issues, will produce new learning in all participants.

Since the beginning of the Forum in 2002, classes of neophyte school leaders enrolled in a school leader preparation program in the United States, and similar classes of students enrolled in a school leader preparation program in Australia, have participated in the Forum. The "global" component of the Forum incorporates these widely separated graduate students who are studying school leadership online from many locations in Australia and South-east Asia, interacting with students based in the mid-west of the U.S., who are also studying school leadership in a predominantly face-to-face mode. Currently, there is potential for participants from Quebec (with English as a second language) and from the UK to also participate.

Using Blackboard as the most readily available communication medium familiar to all participants, and one requiring minimal customization, these learning partners dialog, share resources, collaborate, develop personal and professional relationships, share classroom activities, collaborate on course assignments, argue, reflect, research, and learn together. As they enhance their knowledge, understandings, skills, and dispositions on school leadership issues, their expertise in other, less obvious areas also expands.

These participants learn to understand differences in cultural perspectives, and the subsequent impact on approaches to shared areas of interest. While most school principals are interested in similar school related issues, these Forum participants learn to break the nexus between culturally transmitted and traditional solutions to school leadership issues, and hear how others with different "cultural and professional baggage" come to different solutions for the same problems. Learning to understand alternative conceptions and approaches to the same issues from the perspectives of their new global colleagues frees up the thinking of school leaders in training, and provides them with skills that can be applied in situations where more subtle differences abound.

Further, cultural understandings impact other areas of learning associated with these interactions. While many would consider Australian and American cultures to be quite similar, it becomes obvious, during these interactions, that a more refined understanding of each of the participating cultures is shared by both participant groups.

Other obvious learning benefits result from these interactions. While clearly dependent upon the presence and capabilities of highly interactive information and communications technologies, the Forum actually makes the technology invisible, as the power of the global interaction takes over, and the intensity of the model of learning is activated. Technophobe participants work within a strong peer network of support, and soon lose their focus on the 'technology." At the end of the semester, they are amazed to realize that they have used quite sophisticated interactive technologies, and done so with ease.

In addition to the technology experience, Forum participants also interact with quite a different model of learning. Forum participants collaborate with peers, define directions for dialog, develop responsibility for success of the learning experience, own the learning, ensure authenticity of the interactions, provide evidence that they have achieved the recognized and publicized objectives of the Forum, collaborate with peers and global partners, act as a team, and ensure the quality of the product that is presented to the global audience of which they are part. All of this, supported with scaffolding provided by professors in both locations, but with sufficient "ownership" imposed on learners, by professors, for the value of teaming and collaboration, to dictate the level of success of initial interactions. In-depth monitoring of interactions and levels of ownership allows professors to inject further direction and support for floundering teams, in sufficient time to allow ultimate success, within the bounds of a semester, to be realized.

This approach provides yet another professional choice for them as school leaders, based on their own, personal learning experience, and broadens the scope of the definitions of learning available to this new generation of school leaders. Having experienced for themselves another "form" of learning that for many is new, and for most would have remained outside of their learning experience, these future school leaders take new conceptions of technology-based learning with them. For each school that these Forum participants contribute to in the future, there is a new conception of learning that has become part of the school leaders own learning history, and consequently, becomes an alternative in the choices available when learning is defined in that future school context. The Global Forum on School Leadership displays all of the hallmarks of Type II applications of technology, and presents a strong argument for extending the model to include global relevance, and a focus on twenty-first century learning skills as integral constructs of the definition.

Exploring the Constructs Underlying the Global Forum on School Leadership

This Forum is predicated on the understanding of global learning, as defined by the combination of global reach and global perspectives, to produce a global graduate (Rimmington, 2003). In the case of the GFSL, interactive communication technologies, such as a Web-based course management system (Blackboard), and internet-based videoconferencing capability, provide participants from around the world with the global reach they need to interact and learn together. Global perspectives are added to the mixture when learners interact with other learners living in other countries and cultures. Through involvement in global learning, so defined, the global graduate, or global learner in this case, develops a high level of understanding of and regard, honor, and respect for people of other cultures, and the differences in thinking, world view, and orientation they hold, while becoming more expert in the understandings they have of themselves, and their own discipline or profession.

To place these thoughts in a more contextualized educational arena, school leaders, in the United States for example, are being held increasingly responsible for developing new visions for technology usage for ensuring that recognized standards are being met and surpassed by all learners (ESEA, 2002; Technology Standards For School Administrators (TSSA) Collaborative, 2001). Without providing alternatives to the mental models that new school leaders are often socialized into accepting, leader preparation programs renege on their responsibility to encourage change and continual growth in new generations of school leaders.

It is clear that school leadership in the twenty-first century will require new skills, new knowledge, new behaviors, new dispositions, and new visions. To develop these new orientations effectively, however, future school leaders must experience new learning possibilities in their own education before they can lead others to use a new vision to guide their day-to-day learning practices (Gibson et al., 2003). To this end, Forum participants enhance their comfort with interactive communications technology, expand their understanding of alternative conceptions of common school leadership issues, and increase their own multicultural awareness and sensitivity (Kirkwood, 2001; Pike, 2000). At the same time, and based upon their own reflections and their willingness to incorporate similar experiences in the learning environments for which they are responsible, they experience a new and effective model of learning. In presenting the challenge to tomorrow's school leaders to "develop a deeper, unrestrained, and more sensitive understanding of their areas of responsibility, and in their professional behaviors related to topics of professional concern" (Gibson, 2002a, p. 2), particularly through "involvement in a collegial analysis of views and experiences that vary from the norm or the expected"(Gibson, 2002a, p. 2), the Global Forum encourages both "intellectual and professional challenges to the status quo, [and] to the standard and expected culturally derived response" (Gibson, 2002a, p. 2) to the predominant school leadership issues of the times.

Evaluation of the Effectiveness of the Global Forum on School Leadership

In the current iteration of The Global Forum for School Leadership, school leaders are provided with an international, interactive learning arena designed to provide them with an opportunity to explore, with colleagues from around the globe, issues related to the philosophy and process of leading schools. In a given semester, Australian and American students have the opportunity to interact with each other, and with educators from Canada and the UK, as they explore concepts related to technology, leadership philosophy, curriculum, and learning theory. All of these topics include a heavy emphasis upon the leader's role in these processes, and the type of leader behavior and action that would be necessary in change-oriented and future school contexts.

These participants have left no doubt, through their reflective journals and their contributions to online discussions, about the value of this type of activity for broadening the global horizons of participants, and expanding the learning modalities available to them. Further, because of the unique nature of this learning experience, participants have recognized the need to formalize the evaluation of the Forum, in order to support its continued growth, and to assess the relative learning return on the time invested by all participants. This is an integral component of Forum responsibilities (Gibson, Schiller, & Turk, 2003). Despite the presence of frequent, unsolicited and informal comments supporting the value of the Forum, more formal evaluation of the experience has produced more reflective responses, and more usable feedback, based on the structure and type of the evaluation instruments used. For this learning experiment, four different approaches to evaluation were adopted: a formal survey response form was expected to be completed by all participating students at the conclusion of each Forum (a total of 107 students during three semesters of operation); an analysis of required monthly reflective logs submitted by students was undertaken; usage statistics were generated based on an analysis of participant interaction patterns (a feature of the Blackboard course management software); and an analysis of participant postings to the discussion boards was undertaken to extract reactions related to the operation and value of the experience to participants.

These data sets were able to provide a triangulated description of the operational procedures of the Forum, and the learning benefit accruing to participants. Representative evaluation items, associated data, and a selected range of responses from all semester evaluations are organized below in sections related to forum procedure, technology learning, professional learning, cultural learning, new ways of learning, and overall reactions to the Forum, and presented to provide a sense of the range of reactions to the Forum experience.

Notwithstanding the four approaches taken in evaluating the Forum, the types of questions used in various forms are represented by the following selected items:

- Have you ever participated in this type of online discussion previously? Would you describe your reactions to this experience?

- What advice would you provide to make the Forum more effective?

- What value has the Forum added to your understanding of:

  - school leadership issues?

  - online discussions as a learning tool?

  - the value of adding a "global" perspective to your approach to learning?

  - the need for increasing global sensitivity and cultural awareness?

- What potential do you think the Global Forum on School Leadership has for leadership preparation programs?

- How has adding a "global" perspective to your learning about school leadership assisted you in your growth as a school leader?

- What have been the most valuable "cultural" learnings derived from your participation in the Global Forum?

- What arguments would you make for considering this online process a viable learning tool for your colleagues and your students?

- What have you learned from participating in this process?

While it is clear that much of the data gathered in each of these semesters of Forum operation could be used in a variety of categories for discussion, the following analysis is designed to be representative of the type of responses received following a semester long forum experience. Further, these data are used extensively by forum coordinators in an iterative approach to restructuring and improving the forum experience for succeeding semesters.

Forum procedure. The following comments relate to the type of feedback received related to the operation of the forum, and the problems that resulted for first-timers and experienced participants. Advice to Forum organizers can also be garnered from these responses:

- With anything new, it takes some time to work out all of the kinks. It can become overwhelming with all of the messages.

- It takes a lot of time to read the messages, but I found them to be interesting and exciting. I also felt that I did not have a lot to contribute sometimes, so I found myself more of an observer rather than a participator. I will work on this next time.

- I was disappointed however, that I did not receive a response from the person that I e-mailed personally.

- I often felt that I had nothing to add to what had already been said.

- I think that sometimes they [online discussions] are intimidating as some people may not want to put their thoughts in writing for others to respond to.

- It is more threatening than verbal expression in class.

- It is a good way to include everyone.

- In class, so often a few people dominate discussions. This does provide a level playing field.

- I would say that there should be a certain amount of participation required each week because if you don't, then some people never make a valid contribution.

- I felt overwhelmed at first until we discussed the forum in class and the professor's expectations.

Technology learning. Comments in this category were often difficult to separate from the Forum procedure category. Some difficulties in forum participation could be traced directly back to user comfort with the technology. Many of the technology-related comments also fit nicely into the category related to the Forum as a new way of learning.

- I enjoy the Forum because it will help develop my own skills as a leader as well as help develop my technological skills.

- It has been cumbersome to navigate Blackboard for the extremely short time that I have been accessing it. I'm sure with practice, it would become much easier. I'm still in the forward/backward, check it out stage... like playing Hide 'n Go Seek!

- My only frustration is that it seems that the same people always reply or respond.

- Somehow we need to do a better job of making it easier and making people feel more welcome to share their perspectives!

Professional learning. Comments in this category relate directly to the impact of Forum experience on individual understanding about school leadership and associated issues.

- I think that the forum is a very good way to discuss school leadership issues.

- I actually took a Blackboard training so I could use this method to facilitate discussions with some of my special education teachers.

- I think that the forum provides another means to get to know colleagues.

- I feel that we are on the verge of beginning a global conversation to share our problems as well as our solutions. This can only be good for us all!

- Many of the issues are universal.

Other data extracted from forum postings provided comments exploring leadership issues, captures "gems" of leadership wisdom, and demonstrate the value of the professional exchange represented by the Forum.

- Good leadership will catch people doing things right and acknowledge them for their good work in a positive manner

- Being a leader means being able to listen to others, and being able to consider a subject from all points of view

Cultural learning. In this category, responses relate to a variety of perspectives on the impact of a greater understanding of different cultures.

- I learned that we are all so similar as human beings and we share many commonalities with our overseas colleagues.

- I learned that some Asian parents put a lot of pressure on their children to succeed and don't allow them to be "young."

- One of the most important things we can do is increase global sensitivity and cultural awareness.

- The most valuable "cultural" learning derived from my participation in the Global Forum has been learning that other cultures face similar successes and problems.

Responses to the personal stories of Forum participants provided their own motivation for greater understanding and learning about cultures and people, and the value of the different perspectives they might bring to an issue.

- My family escaped from South Africa during the worst time under apartheid.

- I am one of the "boat people, first generation Vietnamese in the U.S.

- I grew up in Kansas and will spend my whole life here.

New ways of learning. Forum participants displayed a variety of comfort levels with models of learning that varied from the traditional "teacher-centered" model. Few had experience using technology as a tool in their own learning, other than in using word processing or internet searches in order to produce course assignments.

- I would love to have this format up and running for discussions with others who have my same type of position, responsibilities.

- However, I did learn from "listening" to others who are starting out in the field.

- The information helped me reframe some of my thoughts.

- I have learned about a good tool to use in my own learning.

- My own thought process was expanded.

- It is a good way to elicit participation from everyone.

- Forums such as this will become a main component of any class.

- Online discussions as a learning tool? I think the potential is amazing!!!

- It is always good to hear other perspectives...sometimes, that is where our great "AHA" comes from.

- I was able to learn a lot about my fellow classmates, that I did not know and I sit next to some of them each week! It allowed me to get to know them better.

Overall reactions to the Forum. It was clear from the array of responses from all sources of evaluative data that all participants in the Forum benefited in a variety of ways from their participation. These responses have been selected to demonstrate the variety of overall reactions participants had to the experience. The overall success of the Global Forum on School Leadership was confirmed by these data.

- I certainly encourage you to continue this type of a forum.

- I have learned that I am not alone. I have support and I am beginning to set up a network of colleagues.

- This was a great learning experience for me.

- It gave insight to the similarities in educational issues beyond the continent.

- I think it opened my eyes that we are very much a global society and that I should look beyond my own country. I often find it difficult to look beyond my own state!!

Decisions regarding future versions of the Forum will be based upon suggestions, criticisms, and revelations derived from these data. Forum facilitators will redesign the structure of succeeding forum experiences in response to negative comments, and to reduce dysfunctional features. Future topics for the Forum; timing issues related to student work and study life; variations in expectations and scheduling related to semester schedules in universities in the northern and southern hemispheres; issues related to increasing cultural sensitivities, and learning modalities predominated these various forms of feedback. Further, procedural issues related to structuring the analysis of student participation rates, assessing the quality of student postings relative to the awarding of semester grades, and the analysis of the value of and organization of small group activities (particularly when members of the small group were on either side of the equator!) also made this feedback valuable for the learning of the professors involved.

Regardless of the style of evaluation adopted in interactive, international learning community activities, such as the Global Forum on School Leadership, it is clear that participant voices need to be heard and incorporated in future planning decisions. A variety of authentic assessment pieces is recommended, as is variety in the modes through which evaluation is mediated and collected. The ownership and responsibility for success that individual participants displayed during this experience demanded their authentic involvement in the assessment and evaluation of the innovation, and ensured improvement and growth for future versions of the Forum.

Future Trends: Global Forum on Educational Research

With clear evidence of enhanced learning resulting from participation in the GFSL at the Masters level, academic staff in the Department of Educational Leadership have agreed to collaborate on the next generation of the Global Forum. Consequently, the next iteration of the Global Forum Series is the Global Forum on Educational Research (GFER), designed to be an international learning community of graduate students in doctoral study collaborating upon issues related to doctoral research activities. Designed to incorporate (a) an online forum using Blackboard; (b) a series of interactive video, International Research Summits, using existing Polycom video equipment, or Marratech software, where multiple participants from multiple sites will collaborate in multipoint to multipoint Web-based video conferences; and (c) individual to individual, or small group to small group interactive video sessions using computer-based interactive video equipment such as iSight cameras or PC-based fixed-focus laptop cameras, this iteration of the Forum activity has attracted a competitive university grant to support its development.

Topics of analysis will emerge from doctoral student need, and will be tied to progress in doctoral research activities. Coordination of topics will be arranged through a collaborative partnership with professors, researchers, and scholars from each participating institution and designed to tap the research expertise of each institution.

The GFER is initially designed for students enrolled in the doctoral program in educational leadership. It is predicated on the belief that exposure to other views and approaches to research, other cultures, other approaches to learning, and other uses of technology will benefit the growth of the individuals involved in the project.

Further, given that students enrolled in the EdD in Educational Leadership are already employed as full-time leaders in school districts, principals, or superintendents of school districts, a second focus of this Global Forum is to provide these school leaders with authentic and hands-on experience in the design and management of the learning experience through involving them in coordinating and managing the GFER. The belief is that if these school leaders have the experience of contributing to, and participating in, a learning environment suitable for twenty-first century learners in their own learning histories, they will be more likely to transport that experience to their own school systems, and incorporate it into the conceptions of learning, and the approach to learning that predominate in public school systems throughout the country and throughout the world. In this manner, the impact of a Wichita State University (WSU) based global learning experience for school leaders enrolled in doctoral study at this university has the potential to impact other institutions, other professionals, other existing students, and other generations of learners in Kansas ... and in other parts of the world! An additional component of experience is provided participating students who take the opportunity to establish international networks of support related to doctoral research, and follow up with the opportunity to travel to and work with partners at participating institutions to solidify their global experience.

The theoretical and conceptual framework of this project gains support from work in the literature on problem-based learning, global learning, technology-based student-centered learning environments, constructivist teaching principles, adult learning, online learning communities, and the instructional design literature. This GFER project provides direct benefit to students by providing an opportunity for graduate students in Educational Leadership at WSU to

- share responsibility for the design and management of the GFER project;
- participate in online discussions, multipoint interactive video International Research Summits, and individual video conferences related to doctoral research with global colleagues and experts;
- develop expertise in use of Blackboard course management software, multipoint to multipoint video conferencing equipment, and point-to-point computer-based interactive video equipment;
- collaborate with international peers and colleagues on issues of mutual research interest;
- gain exposure to cultures, worldviews, and perspectives different to their own;
- experience learning success through alternative approaches to learning;
- increase exposure to interactive technologies and alternative learning modalities;

- enhance understanding of cultures and the ways of thought of others who are different.

While heavily dependent upon the success of the Global Forum on School Leadership, the innovation in this second iteration of the Global Forum Series focuses on the incorporation of graduate students in the development and management of the GFER as part of their required doctoral coursework and experience. Further, the use of videoconferencing capabilities that facilitate laptop, multipoint conferencing software as the major medium of the interactions, supplemented by other asynchronous, online interactive modalities, adds another level of innovation to the Global Forum series.

Conclusion: Impact of a Global Forum on Participants' Learning Processes ... or ... Type II Learning Products from Global Forum Participation

Throughout the semester of involvement, Forum participants are required to develop artifacts of their global, collaborative learning experiences. Through an analysis of student journals, collaborative team products, and individual weekly contributions to small group and large group discussions, students display increasing cultural and professional sensitivity to how the life experience, language, history, geography, religion, politics, culture, and context of an individual constitutes the basis of alternative perspectives. McKay (1994), and later, Rimmington et al. (2004) explore this process from the perspective of communication theory, and suggest that participants in communications, like those in the Forum, have learnt how to "paint" their cages in ways that allow others to recognize the bars through which messages must be sent and received. The process is not unlike that to which Toffler (1995) refers as a process of learning, unlearning, and relearning in order to adjust to new contexts and to new participants to the communication. Further, it was one intention of this global learning project, to develop the ability to send and receive messages, which facilitated the emergence of self-regulated or autonomous learning (Alagic, Gibson, Doyle, Watters, & Keys, 2004), thereby providing the necessary conditions for co-construction of new knowledge, and recognition of multiple perspectives in the global domain.

In addition, reflective practice was used extensively by participants to refine their thinking and adjust their personal beliefs and attitudes (Gibson & Alagic, 2003; Gibson, Alagic, Haack, Watters, & Rogers, 2003). Co-reflective exchanges between students in large groups, and in smaller more intimate dialog teams, further aided the development of individual skills in recognizing and breaking down the barriers to communication. It was this process that was central to the development of an online learning community, and it was entirely dependent upon student willingness to explore new learning contexts/procedures, develop risk-readiness, trust, and take responsibility for enterprise success. Success of these enterprises also depends upon instructors' ability to be flexible, respond to tone/frequency of engagements, seek ongoing feedback, collaborate with others in the learning partnership, and share ownership/control of the direction of the dialog with participants (Gibson et al., 2003).

As a result of the cognitive apprenticeship (Duncan, 1996) that Forum participants have undergone, a variety of Type II results have accrued. Supported by the presence of interactive computer technologies, the Global Forum has been able to

- incorporate a global orientation to leadership preparation;
- emphasize the use of appropriately selected technologies to achieve expanded course objectives;
- transform the learning model traditionally employed in leader preparation programs;
- provide evidence suggesting that there is a smarter way of thinking about technology use in learning, and that it is represented not by a focus on technology, but a focus on the intent of the learning activity, and the way the learning environment is conceived and structured;
- use technology to support individual growth, professional leader development, and the transformation of commonly accepted cultural stereotypes of organizational behaviors as they relate to leading in educational contexts;
- acknowledge the central role of school leaders in the successful integration of technology into learning environments, and the concomitant transformation of traditional paradigms of learning, pedagogy, and school organization.

The deep thinking, reflection, and the construction of meaning in a global learning environment, as represented by the Global Forum on School Leadership (Gibson, 2002b), presents an example of the currently known upper end of the evolutionary continuum of technology use in learning, as described by Dwyer, Ringstaff, and Sandholtz (1990a). In fact, this process of "instructional evolution in technology" has been described, in their conception, by teachers moving from adopting technology in support of common instructional practices, to adapting technology for experimenting with different instructional practices, to appropriating technology to create new strategies, to creating learning situations where technology is used by students to invent learning experiences. At the upper end of this continuum, teachers who choose to incorporate ubiquitous information and communication technologies into their learning environments have generated a context where learning opportunities themselves are ubiquitous, and conceivably, extend to a global scale. Selby and Pike (2000) do not see this as a choice, however, suggesting that "worldmindedness," as they call it, is no longer a luxury, but a necessity for survival in the new century. They suggest that encountering diverse viewpoints and perspectives engenders a richer understanding of self, concluding that personal discovery is critical to self-fulfillment, and to the generation of constructive change on a global scale. Pike (2000) corroborates through an emphasis upon the interdependence of all people within a global system, and recognizes that within a school learning context, this is more often than not expressed in terms of the connections between students in one country with people and environments in other parts of the world, and the resulting insights, ideas, and information that enable students to look beyond the confines of local and national boundaries in their thinking and aspirations.

Students in our schools today live in an increasingly complex and interrelated world (Calder, 2000). They must share a regional and global responsibility in addition to responsibilities in their own communities. They must be able to take a global perspective, one that challenges injustice, poverty, and destruction, wherever it is found. The quality of life they will have as adults will depend on their ability to think, feel, and act from a different perspective than the one many teachers grew up with. To be effective in dealing with these globally aware learners in their schools, educational leaders must actively recognize that their worlds are changing too, and develop a vision for education where technology applications support deep learning processes (Taylor, 1998), and where they realize that Type II applications of technologies emphasize the learning process and the theory and philosophy behind it, and "are only part of the instructional approach that determines their success" (Maddox et al., 1997, p. 27).

Expectations for education are changing. The knowledge base of education is changing. Conceptions of how individual learning occurs are changing. The tools available to "do" education are changing. The roles of teachers are changing. Understandings of what should be learned, who should be learning, how they should learn, where they should learn, and when they should learn, are changing. So, expecting school leaders to recreate their conceptions of what constitutes appropriate leader behavior should also change! Technology applications, like the Global Forum on School Leadership and the Global Forum on Educational

Research, make available "new and better ways of teaching and learning" (Maddox et al., 1997, p. 22), and represent one way of extending the Type II potential of information and communication technology to new school leaders who are learning their trade in the global learning arena of the twenty-first century.

References

Alagic, M., Gibson, K. L., Doyle, C., Watters, J., & Keys, P. (2004,). The potential for autonomous learning through ICT. *Proceedings of the 15th International Conference of the Society for Information Technology & Teacher Education* (pp. 1679-1684), Atlanta, GA. Association for the Advancement of Computers in Education.

Boud, D. (1985). Problem-based learning in perspective. In D. Boud (Ed.), *Problem-based learning in education for the professions* (pp. 13-18). Sydney: Higher Education Research Society of Australia.

Boud, D., & Feletti, G. (Eds.). (1991). *The challenge of problem-based learning.* London: Kogan Page.

Bridges, E. M. (1992). *Problem-based learning for administrators* (ED No. 347 614). Eugene, OR: ERIC Clearinghouse on Educational Management.

Breck, J. (2004). *Connectivity: The answer to ending ignorance and separation.* Lanham, MD: Scarecrow Education.

Calder, M. (2000). A concern for justice: Teaching using a global perspective in the classroom. *Theory into practice, 39*(2), 81-87.

De Grave, W. S., Boshuizen, H. P. A., & Schmidt, H. G. (1996). Problem-based learning: Cognitive and metacognitive processes during problem analysis. *Instructional Science,* (24), 321-341.

De Vaney, A., Gance, S., & Yan Ma (Eds.). (2000). *Technology and resistance: Digital communications and new coalitions around the world.* New York: Peter Lang.

Duncan, S. L. S. (1996). Cognitive apprenticeship in classroom instruction: Implications for industrial and technical teacher education. *Journal of Industrial Teacher Education. 33*(3), 66-86.

Dwyer, D. C. (1996). The imperative to change our schools. In C. Fisher, D. Dwyer, & K. Yocam (Eds.), *Education and technology: Reflections on computing in classrooms.* San Francisco: Jossey-Bass.

Dwyer, D. C., Ringstaff C., & Sandholtz, J. H. (1990a). *The evolution of teachers' instructional beliefs and practices in high-access-to-technology classrooms*. Paper presented at the meeting of the American Education Research Association, Boston.

Elementary and Secondary Education Act (ESEA). (2002). Retrieved March 18, 2004, from http://www.k12.wa.us/ESEA/default.aspx

Gibson, I. W. (2001). At the intersection of technology and pedagogy: Considering styles of learning and teaching. *The British Journal of Information Technology for Teacher Education, 10*(1&2), 37-63.

Gibson, I. W. (2002a). Masters in Educational Administration—The Global Forum on School Leadership. *Global Learning Development Grant Application* (pp. 1-15). Wichita State University.

Gibson, I. W. (2002b). Developing a Global Forum on School Leadership: Using interactive communications technology to enhance the achievement of learning goals in a school leader preparation program. *Proceedings of the World Conference on Educational Multimedia, Hypermedia & Telecommunications* (pp. 612-612). Denver, CO: Association for the Advancement of Computers in Education.

Gibson, I. W., Schiller, J., & Turk, R. (2003). Evaluating the quality of global learning experiences: Considering the interplay between interactive technology and assessment on an international learning community. *Proceedings of the World Conference on Educational Multimedia, Hypermedia & Telecommunications*, Honolulu, HA (pp. 3093-3096). Association for the Advancement of Computers in Education.

Gibson, I. W., Schiller, J., Turk, R., & Patterson, J. (2003). International, online learning communities: Expanding the learning and technology horizons of new school leaders. *Proceedings of the 14th International Conference of the Society for Information Technology & Teacher Education,* Albuquerque, NM (pp. 231-1234). Association for the Advancement of Computers in Education.

Gibson, K. L., & Alagic, M. (2003, May 16-19). Teacher's reflective practice: Adding a global learning facet. *Proceedings of the International Conference for Innovation in Higher Education* (ICIHE), Kiev, Ukraine.

Gibson, K.L., Alagic, M., Haack, C., Watters, J., & Rogers, G. A. (2003). Using global learning to enhance the preparation of elementary teachers in the teaching of mathematics and scienc*e: What we learned. Proceedings of the 14th International Conference of the Society for Information Technology & Teacher Education,* Albuquerque, NM (pp. 1433-1436). Association for the Advancement of Computers in Eduction

Hannafin, M. J., & Land, S. M. (1997). The foundations and assumptions of technology-enhanced student-centered learning environments. *Instructional Science, 25*(3), 167-202.

Heath, M. J. (1997). Instructional design models for emerging technologies. Paper presented at *the Eighth International Conference of the Society for Information Technology and Teacher Education*, Orlando, FL.

Jukes, I. (2000). *NetSavvy: Building information literacy in the classroom.* Thousand Oaks, CA: Corwin Press.

Kirkwood, T. F. (2001). Our global age requires global education: Clarifying definitional ambiguities. *The Social Studies, 92*(1), 10-15.

Maddox, C. D., Johnson, D.L., & Willis, J. (1997). *Educational computing: Learning with tomorrow's technologies* (2nd ed.). Boston: Allyn & Bacon.

McKay, H. (1994). *Why don't people listen? Solving the communication problem.* Sydney: Pan.

O'Brien, T. C. (1994). Computers in education: A Piagetian perspective. In J.J. Hirschbuhl (Ed.), *Computers in education* (pp. 12-14). Guildford, CT: Dushkin.

Partnership for Twenty-first Century Skills. (2002). *Learning for the twenty-first century.* Washington, DC: Author.

Pike, G. (2000). Global education and national identity: In pursuit of meaning. *Theory into Practice, 39*(2), 64-73.

Rimmington, G. M. (2003). An introduction to global learning. *Proceedings of the 14th International Conference of the Society for Information Technology & Teacher Education,* Albuquerque, NM (pp. 1536-1539). Association for the Advancement of Computers in Education.

Rimmington, G., Gibson, K., Gibson I. W., & Alagic, M. (2004). The cage model of global learning. *Proceedings of the 14th International Conference of the Society for Information Technology & Teacher Education,* Albuquerque, NM (pp. 3027-2032). Association for the Advancement of Computers in Education.

Selby, D., & Pike, G. (2000). Civil global education: relevant learning for the twenty-first century. *Convergence, 33*(1/2), 138-149.

Senge, P. M. (2000). *Schools that learn: A fifth discipline fieldbook for educators, parents, and everyone who cares about education.* New York: Doubleday.

Sewell, D. F. (1990). *New tools for new minds: A cognitive perspective on the use of computers with young children.* New York: St. Martin's.

Taylor, H. E. (1998). How in the world does one teach global education? *Momentum, 29*(3), 16-18.

Technology Standards for School Administrators (TSSA) 2001. Collaborative *Technology Standards for School Administrators*. Retrieved October 18, 2001, from http://cnets.iste.org/tssa/

Toffler, A., & Toffler, H. (1995). *Creating a new civilization*. Atlanta, GA: Turner.

Note

This chapter had its beginnings in an earlier, and much shorter version of some of these thoughts published in a special issue of *Computers in the Schools* (Vol. 22, issue 1/2, pp. 169-182), scheduled for release in 2005.

Chapter V

Enhancing Learning Through 3-D Virtual Environments

Erik Champion, School of ITEE, University of Queensland, Australia

Abstract

We cannot begrudge students their envy in looking at popular films and computer games as major contenders for their spare time. While we as teachers could attempt to fight the popularity of games, I suggest a more useful endeavor would be to attempt to understand both the temptation of games, and to explore whether we could learn from them, in order to engage students and to educate them at the same time. There are still few applicable theories and successful case studies on how we could do this using virtual environments and associated technology (referred to by some as virtual reality, or VR). To help answer the question of "but what can we do about it," I will outline several simplified theories of cultural learning based on interaction, and the experience I gained from employing them in two different virtual environment projects.

Introduction

Today's computer games are vast, powerful, and engaging digital environments. Students buy them, play them, and modify them. With no prompting from teachers, they tear through tutorials and manuals, test new hardware and software, and spend thousands of hours engaged in "hard fun," as well as being immersed in creating characters, animation, sound, and 3-D environments.

Yet in academic research institutes, we see a huge outpouring on the advantages of virtual reality, but little significant educational content. While recent academic literature has criticized the content-poor output of traditional virtual reality research, it has so far been reticent in developing guidelines and practically applicable theories for creating virtual environments that succeed as an engaging medium for entertainment and education.

There are many difficult issues in creating virtual environments: not least is the problem of wrestling with cutting-edge technology. However, I propose the major problem that educators should concentrate on, and help virtual reality researchers with, is understanding how virtual environments, as digital media, relate to how people learn.

To explain that relationship, we need some history on the development of virtual environments, and why, despite being developed in research institutes, they do not immediately lend themselves to teaching. I am going to suggest that a major issue has not been technological constraint, but meaningful content. We lack extensive exploration in choosing and creating interaction that is meaningfully related to specific content, and, for various reasons, we lack impartial evaluation of the projects. Once we can understand that relationship, we can develop an appropriate strategy. That strategy must address the issue of how digital media simulates, augments, or replaces traditional learning through appropriate interaction with meaningful content.

Secondly, we need to ascertain whether entertainment technology, that is, games, offers us more accessible ways of developing virtual learning environments than commercial virtual reality packages. I will suggest that they often do, but that there are several major issues to consider when evaluating and using them. Thirdly, no matter what type of technology we choose, we must have a clear idea of the learning-based goals we hope the participants will reach. Fourthly, once we know the learning-based goals, we will need to scope a different type of interaction that through digital media, simulates, augments, or makes possible those goals. Finally, we need to know if, when, and how these goals are achieved; we have to have an evaluation plan in place.

Missions and Concerns

Until recently, virtual environments have been single-user, with limited ability to interact with the environment. In the rare case where they were multiuser, the interaction possible between participants was limited not just by technical constraints or the desktop personal computer (PC) interface, but also by a lack of thematic relationship between the content and the perceived learning experience. As noted by Johnson (1997), participants often feel they are looking at a computer screen, rather than existing in a real place. Weckström (2004) noted that even when there was a feeling of spatial immersion, of "3-D-ness," the environments were still empty and devoid of apparent purpose.

This thesis began from the fact that, when a group of students were exploring and researching other 'virtual worlds' in order to begin developing Marinetta, they reported that all the worlds seemed empty and hollow, like stage sets. There were neat buildings in these spaces but no sense that these buildings had been built for any real purpose. The students noted that these so called virtual worlds did not seem to be worlds at all, but just architectural spaces that did not give them any feeling of worldliness. (Weckström, 2004, p. 9)

Why this separation of "world" from "architectural space?" In order to visualize spaces, architects have not been worried about social agency, about how participants can relate to each other in virtual environments. They have been concerned about presenting the environment in the best possible light, in order to create impressive fly-throughs.

Even though architecture is about the inhabitation of space, virtual reality, to architects, has been seen as a tool to sell the idea of inhabitation in order to be commissioned to build real buildings. Hence, we should not look to them for the best way on interacting with a virtual interactive world, for they are not interested in building them. I have spoken to several architectural visualization experts about this issue and they agree with me: architects see the technology as an extension of computer aided design and drafting software (CADD or CAD), as a way of presenting objects; they are not (yet) worried about interaction.

I have suggested in previous writings that the way we classify, define, and evaluate virtual environments, needs a major conceptual overhaul, and not just a waiting game for faster computer networks or rendering power (Champion, 2003; Champion & Sekiguchi, 2005). I suggest that why virtual environments have not lived up to their potential as learning environments is due to a variety of reasons, but I also suggest these issues can be addressed.

Learning via Interactive Digital Media

Teaching Issues

Hampered by the continual change and advancement of cutting edge technology, and motivated by the allure of predicting the future, the way people have written about and classified virtual learning environments has been misleading and often unhelpful for the design of learning environments. We do need to create guidelines for immersive learning environments, but citing science fiction, movies, and literature is a dangerous move. Meaningful learning comes from meaningful interaction, rather than from photo-realistic representation of fanciful futuristic situations.

A virtual world has to support the following factors: there has to be a feeling of presence, the environment has to be persistent, it has to support interaction, there has to be a representation of the user and it has to support a feeling of specific worldliness. (Weckström, 2004, p. 38)

To sell books, especially on new and exciting technology, it is very tempting to make outrageous promises. Without mentioning the specific examples, (you can find them by looking for books with "cyberspace" or "virtual reality" in the title from the 1990s), these books trumpeted the paradigm shifting promise of virtual reality. The literature was spearheaded by famous academics who were inspired by science fiction television and films. Typically, these academics had not actually created their own virtual environments via the then available, and to my mind, highly unsuitable technology of proprietary CADD packages, and complex programming languages, such as C/C++, not something learnt overnight!

I would agree that virtual reality technology is often perceived as too expensive, complicated, or time-consuming for teachers. However, recent advances in home entertainment technology (such as in commercial computer games) actually brings the potential of immersive environments as a learning medium closer to the classroom, and to the learning "comfort zone" of today's students.

On the other hand, the actual barrier I have found amongst educationalists is not due to the technology, but to the concept of interactivity. Too many teachers view virtual environments as a digital depository of conventional media: they do not yet see that children can learn from these virtual environments, and that they learn not by reading and reciting, but by exploration and "trial and error" interaction.

Perhaps this has also been due to the success of marketing: attempting to find "virtual" products and case studies on the Web leads to a plethora of digital panoramas, and "virtual media" libraries of HTML and word documents. When I think of virtual reality, I think of 3-D space, and meaningful interaction, not merely two-dimensional pictures, and certainly not files that happen to be available over the Internet.

Designing Places for Learning

Virtual environments are often criticized for evoking "cyberspace" but not "place." In other words, they lack the richness of the associations and encounters that occur in real space (Benedikt, 1991; Coyne, 1999; Johnson, 1997). But what is place? How do we design it? How do we know if this design has worked? More specifically, how we can best create it digitally, in order to enhance learning through interaction, is still vague (Champion & Sekiguchi, 2005).

One would think that virtual heritage environments, being designed to educate us in how people from a previous or distant society saw their world, would encompass and transmit their way of living. Yet writers such as Mosaker (2001) and Gillings (2002) have stated that virtual environments lack meaningful content, and virtual heritage environments and virtual archaeology are a case in point. "VR systems do not offer an alternative 'reality'; they do, however, provide simulated worlds that seem 'realistic'" (Schroeder, 1996, p. 15).

Designers may use this conflation to persuade the viewer that high-resolution images imply a high degree of archaeological certainty, when this is not the case.

The distinctions between real and hypothetical are not simple but subtle, complex, and far-reaching .... As Mr. Emele pointed out in his article, a partially known site cannot be reconstructed satisfactorily .... Our reconstructions are also too clean and neat. (Eiteljorg, 1998, p. 2)

Being designed for visualization, rather than for interaction, most virtual environments are single-user. Where they allow several people to see each other, sharing of information is usually restricted to chat, sending files or hyperlinks; control of social interaction is limited. People are, by nature, social creatures: they will almost certainly want to interact with and be recognizable to other participants. On the other hand, they might want some control over the quantity or even quality of social interaction.

In order to understand how we learn through interaction with the environment, I turned to the literature of cultural geography. Here I found the ideas of Relph, in his book *Place and Placelessness,* offered me a way of both *describing* and *prescribing* attributes of virtual environments.

The identity of a place is comprised of three interrelated components, each irreducible to the other, physical features or appearance, observable activities and functions, and meanings or symbols. (Relph, 1976, p. 61)

Relph's book enabled me to see that virtual environments are typically designed to aid visualization, to foster activities (such as games), or (potentially) allow people to develop their own projected identities, and interpret that of others, either directly via social interaction, or indirectly through observing, investigating, and through playing with cultural practices and artifacts. And it is easiest to create the first type of environment, for it involves creating an interface that facilitates navigation around static objects, but no other *directed* form of interaction. The second type, the activity based virtual environment, requires the creation of goals, strategies, and, (typically), rewards.

Collaborative learning environments typically require the notion of an *owned* and *shared* space (or place). While the first two types of virtual environments do not require collaboration, the third type typically does. As Yi-Fu Tuan notes (1998), culture adds a nonvisible layer of interpretation to visible objects, yet virtual environments typically only attempt to simulate what is there.

The third type of virtual environment must somehow present the notion of the intangible; it is genuinely difficult to create. The third type of virtual environment requires the environment itself to be both modifiable and readable in a meaningful way. This third type of environment is the most desirable learning environment: rather than learn about objects (produced by people), or procedures (learnt via activity), the most powerful way of learning is to see how people identify, share, and commemorate what is important to them.

Hence, to foster this form of cultural learning, we require interaction. The approach suggested here is constructivist; learners construct their own meaning (Hein, 1991). In his article, Hein argued that interactivity in exhibits creates more engagement by allowing the user to apply the tool directly to their own life. I totally agree with him. In my own evaluations of people using virtual environments, the most popular requests have been for personalization of their avatars, and for modification of the environment.

For a learning environment, social agency allows students to motivate each other through competition or collaboration. It allows the teacher to "enter" the "world" of the user as a fellow participant rather than as an invigilator. Social agency can help structure and direct the learning experience. Coupled with personalization, social agency encourages the student to explore, take control, and express themselves to others (which requires they learn the skills to do so).

Secondly, the WIMP (windows, icons, menus and pointers) interface of typical workstation computers can be cheaply replaced, thanks again to the aggressive marketing and add-ons of gaming technology. There are force-field generators, exercise machines connected to console games as navigation and interface devices, "VR goggles," skateboards, haptic devices, and even low-cost biosensors that control navigation and selection of objects through monitoring the participants stress levels and heartbeat. In all these cases, the products are available for a few hundred American dollars, not the thousands of dollars of specialized commercial virtual reality solutions.

Nonconventional Learning Suitable for Virtual Environments

As a learning platform, the virtual environment, and its much-hyped sister, Internet-based distance learning, has been savagely attacked by philosophers such as Dreyfus (2001). Critics have argued that virtual environments cause disembodiment, disorientation, discomfort, and social alienation.

On the other hand, psychologists use virtual environments for curing phobias (especially spatial ones), distracting the attention of the participant during painful surgery, improving hand-eye coordination, for leadership simulation training, and even for improving cultural understanding of soldiers in foreign lands.

Virtual reality therapy already helps many patients overcome phobias: from fear of flying to fear of spiders. Jacobson (2000) notes similar systems are being tested to see if they can reduce bouts of anorexia and bulimia.

Virtual environments can help promote technology for the sake of technology, for example, in creating 3-D product showcases, available over the Internet. However, this does not help teachers create learning environments: being impressed by technology is not the same as being inspired by it.

More importantly for teachers, digital media can synergise learning by the use of various multimedia, for example, 3-D modelling packages can be used to create models of the human heart in action, or they can use graphic cutaways, 3-D models, and sound, to demonstrate how to service a car's engine. The same technology can also preserve cultural artefacts through a three- or even four-dimensional record of history.

The interactive possibilities include seeing a reconstruction, as well as an idea of how it was inhabited, along with artefacts used as they had been intended, and in context. Time-based media can present ideas, objects, or techniques that are difficult to visualise either in real life, or through conventional media. Yet imaginative digital visualisations and reconstructions are only part of the story. Even more importantly, through interactive digital media, we can learn by doing.

Solutions

Interaction for Meaningful Learning Experiences

In education circles, it is possible to find Websites with two-dimensional information labelled "virtual" or "digital" media. But to do so is to miss out on the magic of three-dimensional space, which is necessary to create a rich and atmospheric sense of spatial presence. Coupled with meaningful activity, space becomes place; it identifies, commemorates, and records the meeting of a group with a task.

"Yet, paradoxically, remote learning — one of the rapidly growing uses of Cyberspace — is little more than an organized way of distributing learning materials in an efficient, electronic way. All the participants in 'remote learning' (the teachers and the students), miss out on the rich, cultural and social phenomenon of the learning experience itself. While the method is efficient, it is hardly the type of experience most learners will remember fondly many years later, like remembering their place-specific high school or college years" (Kalay, 2004, p. 196).

Once we have three-dimensional space, students can more easily see what can take place inside the environment. They can learn about spatial proximity, they can personalise the environment in relation to others, and identify themselves to others through the creation of avatars and by the "marking" or annotation of the virtual environment.

For this process to be meaningful, and for there to be enough motivation for individuals to meet and collaborate, there needs to be activities and goals that allow the participants to explore, attempt, and to identify themselves with or against. We also need to have clear ideas as to how types of interaction affect the ways in which we learn, and how they can be approximated by the limited interaction typically available in virtual environments.

What Games Have to Teach us about Interaction

Making content appealing to the end-learner may be the lesson that the e-learning industry needs to learn most of all. (Aldrich, 2004, p. 7)

A considerable amount of literature has argued that interactive engagement in a computer medium is best demonstrated by games (Aldrich, 2004; Champion, 2003; Laird, 2001; Manninen, 2004; Prensky, 2001; Schroeder, 1996).

Constructivism and constructionism are education theories that seem to directly support the use of games as learning environments (Brooks & Brooks, 1999; Wong, 2003). For example, Papert worked with Alan Kay on the cross-platform and open source project Croquet. Croquet is designed to allow participants to meet each other in 3-D spaces, which allowed them to collaborate on any files on any PC, even if some of the participants did not have the native applications to open them (Lombardi & Lombardi, 2005). Perhaps more radically, Prensky argues that students of today perceive and think differently to past generations (Prensky, 2001). He believes that by using their cultural artifacts to communicate to them, we are both acknowledging their cultural worth, and are more likely to impart learning that is both more accessible and more meaningful to them.

Some teachers use commercial game engines and online role-playing games as a catalyst to talk about social identities (Gee, 2003); others use games in class to teach students historical processes and how to examine counterfactual history (Squire, 2002). Games can also be developed to enhance and discuss collaboration and teamwork practices (Squire, Makinster, Barnett, Barab, & Barab, 2003). Designing an engaging virtual environment, which challenges, fosters skills, and inspires new learning, in order to develop successful strategies, can thus be helped by an understanding of game design. Games can have context (user-based tasks), navigation reminders, inventories, records of interaction history (i.e., damage to surroundings), and social agency. Games are a familiar medium to users, and help train us how to learn and how to use props as cultural tools. Games provide competition, and therefore challenge: a feature typically lacking in virtual environments. Further, just as the most popular games (excluding Tetris) require representations of opponents (social agents), so too do virtual environments. As in games, virtual environment users may prefer personalization. Engaging virtual environments also requires interaction geared towards a task, a goal. This is a crucial feature for learning environments: the virtual environment as game has to motivate the participant, not the teacher (Amory, Naicker, Vincent, & Adams, 1999).

As users become engaged in the tasks, it is easier to observe them without damaging their level of engagement, especially as games traditionally have built-in evaluation mechanisms. Furthermore, games cater to learning curves of new users by advancing in complexity over time, and this can be incorporated into virtual learning environments.

If we could only crack the issue of why the "hard fun" of games has not been seen in institutional learning, we may be able to create *educational* and *engaging* learning experiences, using the techniques of these highly immersive virtual environments, and matching them to meaningful learning goals.

So what is a game? Part of the attraction of games is certainly due to their interactive and engaging nature, as explained by the following definition by

Aldrich (2004, p. 240). For him, a game is "An interactive and entertaining source of play, sometimes used to learn a lesson."

More helpful for designers is the definition by Salen and Zimmerman, as it attempts to explain what makes games entertaining. In their large tome on game design, they wrote the following, often-quoted definition of a game. "A game is a system in which players engage in an artificial conflict, defined by rules, that results in a quantifiable outcome" (Salen & Zimmerman, 2003, p. 572).

To my mind, one powerful feature of games has been downplayed in the above example. Games are systematic in that they are rules-based, but they are also characterised by inspirational difficulties. The harder the tasks in a game, the more people are inspired to try to solve them. With that in mind, here is my working definition of a computer game (different to Salen and Zimmerman); a game is a challenge that offers up the possibility of temporary or permanent tactical resolution without harmful outcomes to the real-world situation of the participant.

Games Culture and Learning

Many people may well object that games merely involve developing hand-eye coordination, that they do not have meaningful interaction, for people do not learn about other cultures and societies. In order to answer this important question, I wish to reconsider the nature of culture, and how we can develop a model of how culture is learnt.

"Culture consists of patterns, explicit and implicit, of and for behaviour acquired and transmitted by symbols, constituting the distinctive achievement of human groups, including their embodiment in artefacts; the essential core of culture consists of traditional (i.e. historically derived and selected) ideas and especially their attached values; culture systems may on the one hand, be considered as products of action, on the other as conditioning elements of further action" (Kroeber & Kluckhohn, 1952, p. 357).

Given the above definition of culture, we can see that cultural learning involves different modes of interaction. When we visit other cultures, we often learn new, and hitherto foreign, cultural perspectives through copying others' behaviour. We also learn through listening (to their language, to their myths and music), or through reading text and viewing media (as tourists and students). However, we also learn by making hypotheses, by applying the right words in the wrong situations, and learning the correct phrase or protocol through social embarrassment and sometimes ridicule.

Consider, for a moment, the distinction between travellers and tourists. Ideally, we would also try to develop interactive scenarios for inhabitants. Unlike tourists and even travellers, inhabitants have certain roles, responsibilities, and powers (Gee, 2003; Weckström, 2004).

Travellers require more contextual interaction than tourists do: they have goals; places to see within a certain period of time, and without too much exertion; people to find or to avoid; items to seek out, purchase, or utilize. Inhabitants are more constrained, yet knowledgeable in different ways than visitors (travellers and tourists); they have certain place-related tasks to complete using local resources. Giving people goals (as travellers or inhabitants) may increase their engagement or sense of authenticity (Mosaker, 2001; for an alternative viewpoint, see Hedman, 2001).

Travellers and tourists learn about places by going there, observing events, and being instructed by signs, by guides, and by reading printed material. Such learning in games is typically before the start of the actual game, in the form of a textual introduction or voice-over. It is seldom part of the overall gameplay itself. Game tutorials, on the other hand, are procedural. Some games offer short walkthroughs, where users may practice learning to jump, sidestep, use weapons, and so forth. As contextually appropriate simulations, these games within games offer something real-world tourism seldom encompasses; learning by doing.

When we develop cultural knowledge, we do so through observation, through being told by others, or through trial and error. It is possible to classify game devices (to create challenge or provide information) according to these learning modes.

For example, gamers can learn about the background or potential dangers in the place by how old or worn it is. They also learn as spectators, watching other players, or observing other players strategies while competing with them. Observation-based learning is common to many types of games, and perhaps most evident in Tetris and Space Invaders (they do not require instruction to learn how to play or the outcomes).

Other ways of learning include social learning (by people telling you or instructing you). However, being told what to do is anathema to gamers, as they typically want to act rather than to listen. Games that offer ways of learning how to play without reading instruction manuals seem to be more popular than other games, however, strategy manuals seem popular for gamers to establish themselves as experts once they have mastered beginning gameplay. Ways of providing instructions in games include cut-scenes, dialogue by non-playing characters (NPC), notes found in the scene, the introduction, and strategically placed or timed voice-overs.

The third major way of learning appears to be by trial and error. We can learn about a place through task-based activity there (for example, we learn a swimming pool is suitable for swimming). Gamers learn how to use weapons by making mistakes and firing when too close to a wall. By making mistakes, gamers can learn what is forbidden or promoted by the game rules. For example, in some early shooter games, gamers discovered that shooting missiles at their own feet did not damage them, but allowed them to levitate over walls. This strategy was not foreseen by the game designers, but was later incorporated into the artificial intelligence of the computer-driven players.

I note here that in both computer games and in real life, we often learn through a hybrid of the above. We observe or read why or how people do things, we get some advice on what we are doing wrong, or we overhear how or why other people do it, then we try out different strategies in order to most enjoy it, or most successfully complete the task. Even simplistic 3-D shooter games often involve instruction (at the start), observation (where enemies are hiding or where "health" can be restored), and trial and error (learning how and when to use tools or weapons).

Once we have categorised types of learning, we notice a further separation, between learning by doing, and learning by being shown. Games incorporate both, but "gameplay," the innate and unique engagement created by the game as an interactive goal-directed experience, is built on learning by doing. And it is revealing that learning by doing is the main component of games, wherever possible: learning by being shown is relegated to the beginning of the game, to tutorials, and to online help documentation.

In other words, knowledge developed can be either procedural or prescriptive. The hugely successful market of computer games has shown us that unlike traditional media, interactive virtual environments can be highly useful for procedural learning.

Learning Content that Needs to be Incorporated

I have given several examples of virtual environments that matched content with learning aims and technology. I do not suggest commercial games are directly suitable for education, as they improve hand-eye coordination and spatial cognition: nor are they necessarily suitable for teamwork, or understanding different social perspectives. However, when modified, their cheap and accessible technology, along with the familiarity of their interface conventions to students, can provide for rich learning environments.

There is a developing hybrid game genre that may also prove advantageous to teachers. The new editors for online games and collaborative environments are powerful and inexpensive. The environments that they can support range from role-playing to strategy; some incorporate both. The interactivity can be geared towards collaboration and reconfiguration, rather than outright destruction and competition, depending on the interests of the builder.

Placing the Teacher in the Learning Experience

Where does the teacher stand in relation to the virtual environment experience? Does the teacher become a participant, a level designer, a teacher of the technology? Does the teacher stand back, and help fix things (a one-person help desk), engage with the others in the environment as a character, or evaluate, either remotely or in the classroom?

There are many possible answers, but here I would like to briefly outline a few advantages and disadvantages to each option. For exploring visualization-based archaeology type worlds, leading the class through a virtual environment by voice while allowing each person to navigate their own avatar on their own computer, is useful to the students, and requires less hands-on work However, it is more difficult to evaluate student learning, and ensure that each student is maximizing their learning experience.

For language learning, there still seems to be resistance to using spatial environments. While virtual environment technology can allow students from different cultures, who are learning each other's language, to learn directly from each other, generally, the environments are used as meet and greet 3-D add-ons to chat programs. There is much exciting work to be done in how the reconfiguration of avatar and designed worlds can allow students to identity and test out distinctive social, cultural, and linguistic purposes, but the immediate challenge is, perhaps, to show language teachers what could be done, and make the technology more robust, cross-platform, and modifiable for distance learning.

How We Can Evaluate User Responses

Academic virtual environment evaluation usually involves requesting test users to fill out questionnaires indicating a level of presence against three, four, or five general criteria (a feeling of physical space, negative feelings, social agency, naturalism or realism, and engagement).

Questionnaires are prone to error according to Slater (1999). Evaluating people after their experience of the virtual environment may be prone to error, as it relies on memory recall, and on their noticing and communicating exactly what made their sense of engagement seem powerful, weak, or nonexistent.

If a virtual environment seems "natural" to viewers, they may not notice important features that a trained expert would consider distracting or ineffective. We need "passive" evaluation mechanisms to determine the level and type of engagement, without breaking that level of engagement.

Games are actually highly efficient evaluation mechanisms. Their speed and accuracy in evaluating success, or otherwise, is one of the important features of computer games. Game-style abstraction can be just as engaging to users as a sense of realism. Further, as users become engaged in the tasks, it is easier to observe them without damaging their level of engagement, especially as games traditionally have built-in evaluation mechanisms. However, there are issues in applying evaluation techniques to learning about culture. For learning environments, we are not normally interested in task performance: we are interested in understanding.

If we can define understanding to the point where we can test it, there are many ways that evaluation can be built into a virtual environment, with or without the direct participation of the student. If the environment is goal-based, we can evaluate the student's task performance. If the student can act as world builder, we may be able to incorporate peer feedback, especially if the virtual environment is amenable to annotation by visitors.

In the near future, it may also be possible to gauge the physiological state of the participants. I am not here referring to brain scanning, which is already used to evaluate a sense of virtual presence, but is expensive and inaccessible to most teachers. The introduction of biosensors to commercial games is opening up the market to future innate, automatic, and thematically related evaluation of user engagement.

How We Can Interactively Augment the VE Content

There are several famous projects using augmented reality as learning experiences (Azuma, 1997). However, we can also incorporate real-time data into the game, especially if we are connected to the Internet, or if we are using certain types of game accessories. Why would we want to do that, and is it feasibly within the scope of teachers?

Firstly, I believe it is possible. Digital technology can integrate the real and the conjectural, as well as synchronous and asynchronous data, into conceptual, user-specific information. This capability suggests that virtual environments may augment, for example, real-world travel and tourism experiences, rather than merely emulate them.

It is possible to have various aspects of the environment dependent on real world data, connecting across the Internet. One virtual heritage project contained animated fireflies; their movement was directly dependent on real-time stock market movement delivered via the Internet (Refsland, Ojika, & Berry, 2000). Using a bit of imagination, and a Web-based data mining script, one could show people the effect of tourism, or how changing conditions affect environments.

For example, real-world places have changing conditions: rain, wind, heat, and so forth. Using real-time data that feeds into a virtual environment via a Web-connected database, visitors in a virtual world could see real-time weather conditions of a place on the other side of the planet. There are already games that, via the Internet, download your local weather conditions into the game itself.

Case Study

In this section, I wish to outline two learning environments recently undertaken. The first project attempted to answer the following questions:

1. **Place vs. cyberspace.** What creates a sensation of place (as a cultural site) in a virtual environment, in contradistinction to a sensation of a virtual environment as a collection of objects and spaces?

2. **Cultural presence vs. social presence and presence.** Which factors help immerse people spatially and thematically into a cultural learning experience?

3. **Realism vs. interpretation.** Does an attempt to perfect fidelity to sources, and to realism, improve or hinder the cultural learning experience?

4. **Education vs. entertainment.** Does an attempt to make the experience engaging improve or hinder the cultural learning experience?

The prototype solution was the creation of an online reconstruction of an ancient Mayan city in Mexico, and by using techniques learnt from game-style interaction, to evaluate the effect of certain types of interaction on the cultural understanding and subjective preferences of the users.

I had identified certain devices that I believed aided cultural learning in terms of observation, trial and error, or by conversation. In order to test cultural understanding, I created a virtual reconstruction of a world heritage site (Figure 1). I split the reconstruction into three thematic parts, and then assigned each the three different types of interaction "modes," but with similar content to each other. These interaction modes were instruction (scripted chat agents), observation (finding and reading hidden inscriptions), and activity (having to identify and move objects around in order to navigate through to specific goals). The match of environment to interaction mode, I termed a "world." I then set up tasks, and five different types of evaluation. I also had three different audience groups: archaeology students, visualization and heritage experts, and employees of a major travel publication company.

From the observations of the participants, and by comparing their engagement in these archaeological worlds compared to more game-style environments that

Figure 1. Case study 1

I designed for them, I found that when they were told something was a game, they automatically seemed more comfortable. And although the game-worlds did not score as well on their subjective preference rankings when answering a questionnaire, they were much more unwilling to leave these worlds than the archaeological ones!

With the archaeology students, rather than set them specific tasks, in one class I walked them through the world. That is, I talked about interesting things on the site, and suggested which "view" button related to what I talked about. They controlled the navigation, appearance, and orientation of their avatar, and followed me quite happily through the site, at their own pace. While talking about the history of the site, I asked them how easy it was to navigate, what they wanted to do, and other simple questions. The most popular request was not for more detail and information, but for more ability to change their avatar, and destroy things!

The practical experience I gained from these experiments was invaluable, but not immediate. I say not immediate because the evaluations themselves were not conclusive. I found that gender, age, and computer experience (in descending order) were important factors, but that task performance could not be directly related to understanding (tested by asking general knowledge questions afterwards), as people were divided on wanting to explore or to compete.

After some more questioning, I came to the following conclusions. There were at least three types of participants: those that wanted to explore, those that wanted to chat, and those that wanted to compete. I later found that games researchers suggest there are actually four types of personality profiles (Bartle, 1999).

I also realized that my theoretical model of cultural learning was inaccurate: people do not just learn by observation, instruction, or by trial and error, they also learn through a combination of these methods. I also realized that traditional learning of historical sources is not procedural; it is not learnt by trial and error, but by books and dictation. Many students have difficulty in prescriptive learning, so perhaps procedural learning may offer specific advantages to them. This raises the issue of whether virtual environments should scope out the preferred learning style of the participant and cater to that style.

The game genre is both a powerful device for improving usability (the conventions for interaction, navigation, and defining goals are well known), and a dangerous one. When playing a game, people do not notice anything in the environment that they do not consider directly related to solving the game goals.

We also lack examples of digital interaction in games that are not destructive, and do not relate to user response times. While strategy games involve the learning of resources, they, along with other games, may confuse the participant as to what is real, and what is imaginative. That is, playing a historical strategy game may help one develop an idea of where to go to buy a catapult, or even what the catapults were called, but one can't say for certain if they were used, where they were used, or their symbolic and cultural value to the local inhabitants.

The second project used similar technology to create a learning environment for cross-cultural language learning, with social collaboration between language students in Japan and in Australia (Champion & Sekiguchi, 2005). This was a very interesting project, as many of the staff had initial reservations about why 3-D added to the learning experience. It was also a testbed for new ideas on how certain types of interaction in space can help our learning of such things as foreign languages.

The evaluation of this project is ongoing, but it raised interesting issues of evaluation without direct teacher input, as the students could participate in their own time, and wherever there was a suitable computer that was connected to the Internet. In the first stage, I scripted triggers that recorded who spoke and to whom, what rooms they went into, and for how long. For the second stage, I suggested a review of what learning was actually possible, and whether we could evaluate that directly, rather than merely record where students went, and so forth.

This ongoing collaboration helped me see the issues that nontechnology focused teachers may have in incorporating digital media. The students find it highly engaging, other staff members may fail to see pedagogical returns, and it can be difficult to scope out interactive scenarios that directly relate to learning outcomes. I believe that in this example, we can borrow from usability methods, particularly the "teaching out loud" method. Instead, we could develop the "speaking/chatting out loud" method, whereby students can teach each other their language through asking and answering questions, through trying to work out who is a native or non-native speaker, and so forth.

I have developed strategies for single-user learning through puzzle games, and through scripts that allow for exact, or near exact answers in different languages to open doors, change worlds, or change avatars. However, I believe that getting the students to build and invigilate the learning environments is the most powerful way of using interactive digital media. There should not be a blind push to use digital media as hoops: we should investigate whether world-building and other unique features of virtual environments offer greater educational advantages. We should also reexamine learning itself, in context. Would imaginative environments help real-world learning? Are particular forms of interaction more suitable than others? Can we develop alternatives to computer shooters that engage the students?

Future Trends

In game design, there appears to be a trend towards online collaborative and mixed genre environments. The game level editor as a world builder phenomenon means that designing virtual environments is within the reach of classes. The proliferation of software that allows designers to change between different modelling and animation formats is also encouraging. However, and at the risk of sounding repetitive, there still needs to be more research on meaningful interaction.

The case studies I have mentioned have given me new learning models that encourage student participation: encourage them to separate fact from fiction, allow for innate behind-the-scenes evaluation, and provide for peer feedback, which I believe is an additional powerful form of learning. Unfortunately, I am still scoping out these projects, and final results are not currently available.

"There is a shortage of research integrating theory and practice on how best to augment or invoke the user-experience of place via digital media" (Gillings, 2002, p. 17). By concentrating on achieving photo-realism rather than on understanding the unique capabilities for digital media to enrich the user-experience, there are significant questions still to be answered.

Case studies of learning via game-style simulations exist (Aldrich, 2004), as well as descriptions of how we learn via video games (Gee, 2003), so it seems only a matter of time before performance evaluation can be conducted contextually and indirectly.

Conclusion

I don't believe that virtual learning environments are a waste of money, or an ineffective learning tool. I have taught games design by getting the students to build game levels using the applications discussed, and then getting them to mark each other's creations. Not only did they learn the difficulties of designing for others, but they also began to develop a critical vocabulary relevant to them and to their peer group.

Games are far more accessible, and in some cases, more powerful than specialist "virtual reality" environments, and they now offer editors that individuals can use at home on personal computers to extend or create digital environments that they can then explore and share with others. They also offer inbuilt evaluation mechanisms, and as they are typically goal-based, they offer a platform for creating learning tasks.

The many dedicated and hybrid game genres each afford different types of learning experiences, and we are only just beginning to understand which type of interface and interaction suits specific learning requirements. Briefly, these "worlds" can help augment learning in terms of placing historical studies and events, learning about cultural strategies and processes through trial and error, or through collaborating with students from different social and cultural backgrounds far away. The case studies mentioned also revealed how designing and evaluating cultural environments can, in turn, reveal to us new insights into cultural learning and understanding.

How to challenge students to learn meaningful content is a fundamental, pedagogical issue. Developing game-like interaction seems a promising way of achieving this, for games are challenging, in the sense of being difficult in a good *and* engaging way (Rieber, 1996). The trend in commercial games towards multiplayer, and highly customizable environments, is a positive move for education. For rather than try to get students to learn about static content, which needs to be somehow impressive and dynamic to keep their interest, the new games encourage them to create and communicate meaningful content, rather than just be passive consumers.

It is tempting to suggest virtual learning environments have failed because teachers have not understood the technology. Is the technology too difficult? Perhaps for some of us, but for students the technology is *challenging*. While the tools of creating new game "levels" and game "mods" may appear difficult to teachers, why then are there so many game levels built by students?

I would rather suggest that we, as educators, have not yet fully realized that learning itself, needs to be reexamined. If we see games as user-directed virtual environments, it may become clearer how difficult it is to teach prescriptive knowledge, rather than to allow students to learn procedurally.

The implications of digital media and immersive virtual places are not just novel and entertaining, they may be more suited to the ways in which today's students can learn and express themselves. We need to move past what students appear to be doing when they play games, and see what they are learning, and how they are learning, inside these games.

References

Aldrich, C. (2004). *Simulations and the future of learning: An innovative (and perhaps revolutionary) approach to e-learning.* San Francisco: Jossey-Bass.

Amory, A., Naicker, K., Vincent, J., & Adams, C. (1999). The use of computer games as an educational tool: Identification of appropriate game types and game elements. *British Journal of Educational Technology, 30*(4), 311-321. Retrieved January 7, 2005, from http://www.nu.ac.za/biology/staff/amory/bjet30.rtf

Azuma, R. (1997). A survey of augmented reality. *Presence: Teleoperators and Virtual Environments,* 6(4), 355-385. Retrieved August 9, 2005, from http://www.cs.unc.edu/~azuma/ARpresence.pdf.

Bartle, R. (1999). *Hearts, clubs, diamonds, spades: Players who suit Muds.* Retrieved August 5, 2005, from http://www.mud.co.uk/richard/hcds.htm

Benedikt, M. (1991). *Cyberspace: First steps.* Cambridge, MA: MIT Press.

Brooks, J. G., & Brooks, M. G. (1999). *In search of understanding: The case for constructivist classrooms.* Alexandria, VA, USA: ASCD - Association for Supervision and Curriculum Development.

Champion, E. (2003). Applying game design theory to virtual heritage environments. In S. Spencer (Ed.), *Proceedings of Graphite International Conference on Computer Graphics and Interactive Techniques in Australasia and South East Asia* (pp. 273-274). Melbourne, Australia: ACM SIGGRAPH.

Champion, E., & Sekiguchi, S. (2005). Suggestions for new features to support collaborative learning in virtual worlds. In T. Sakai, K. Tanaka, K. Rose, H. Kita, T. Jozen, & H. Takada (Eds.), *Proceedings of C5: The Third International Conference on Creating, Connecting and Collaborating through Computing* (pp. 129-136). Shiran Kaikan, Kyoto University, Japan: Kyoto University. Retrieved August 8, 2005, from http://doi.ieeecomputersociety.org/10.1109/C5.2005.25

Coyne, R. (1999). *Technoromanticism: Digital narrative, holism, and the romance of the real.* Cambridge, MA: The MIT Press.

Dreyfus, H. (2001). *On the Internet.* London: Routledge.

Eiteljorg, H. (1998*).* Photorealistic visualizations may be too good. *CSA Newsletter, XI*(2). Retrieved January 5, 2005, from http://www.csanet.org/newsletter/fall98/nlf9804.html.

Gee, J. P. (2003). *What video games have to teach us about learning and literacy.* New York: Palgrave Macmillan.

Gillings, M. (2002). Virtual archaeologies and the hyper-real. In P. Fisher & D. Unwin (Eds.), *Virtual reality in geography* (pp. 17-32). London; New York: Taylor & Francis.

Hedman, A. (2001). *Visitor orientation: Human-computer interaction in digital places.* Unpublished master's thesis, The Royal Institute of Technology in Stockholm (KTH), Sweden. Retrieved January 7, 2005, from http://cid.nada.kth.se/pdf/cid_107.pdf

Hein, G. E. (1991, October). *Constructivist learning.* Theory, Institute for Enquiry. Presented at The Museum and the Needs of People CECA (International Committee of Museum Educators) Conference, Jerusalem, Israel. Retrieved January 7, 2005, from http://www.exploratorium.edu/IFI/resources/constructivistlearning.html

Jacobson, L. (2000, August 14). A virtual class act technology aims to help hyperactive students. *Washington Post.* Retrieved October 29, 2003, from http://www.washingtonpost.com/ac2/wp-dyn?pagename=article&node=&contentId=A21148-2000Aug13&notFound=true.

Johnson, S. (1997). *Interface culture: How new technology transforms the way we create and communicate.* San Francisco: Harper Edge, New York: Basic Books.

Jones, J. G. (2004). 3-D online distributed learning environments: An old concept with a new twist. *Proceedings of the Society for Information Technology and Teacher Education* (pp. 507-512). Atlanta, GA: SITE. Retrieved August 7, 2005, from http://courseweb.unt.edu/gjones/wp0304a.html

Kalay, Y. E. (2004). Virtual learning environments, *ITcon, Special Issue ICT Supported Learning in Architecture and Civil Engineering, 9*, 195-207. Retrieved August 5, 2005, from http://www.itcon.org/2004/13

Kroeber, A., & Kluckhohn, C. (1952). *Culture: A critical review of concepts and definitions.* New York: Vintage Books.

Laird, J. E. (2001). Using computer games to develop advanced AI. *Computer, 34*(7), 70-75. Retrieved January 7, 2005, from http://ai.eecs.umich.edu/people/laird/papers/Computer01.pdf

Lombardi, M. M., & Lombardi, J. (2005). Croquet learning environments: Extending the value of campus life into the online experience. In T. Sakai, K. Tanaka, K. Rose, H. Kita, T. Jozen, & H. Takada (Eds.), *Proceedings of C5: The Third International Conference on Creating, Connecting and Collaborating through Computing* (pp. 137-144). Shiran Kaikan, Kyoto University, Japan: Kyoto University. Retrieved August 8, 2005, from http://doi.ieeecomputersociety.org/10.1109/C5.2005.25

Manninen, T. (2004). *Rich interaction model for game and virtual environment design.* Unpublished doctoral dissertation, University of Oulu: Finland. Retrieved January 7, 2005, from http://herkules.oulu.fi/isbn9514272544/isbn9514272544.pdf

Mosaker, L. (2001). Visualizing historical knowledge using VR technology. *Digital Creativity S&Z, 12*(1), 15-26.

Prensky, M. (2001). *Digital game-based learning.* New York: McGraw-Hill.

Refsland, S., Ojika, T., & Berry, R. (2000). The living virtual Kinka Kuji Temple: A dynamic environment. *IEEE Multimedia, 7*(2), 65-67. Retrieved August 8, 2005, from http://portal.acm.org/citation.cfm?id=614975

Relph, E. C. (1976). *Place and placelessness.* London: Pion Ltd.

Rieber, L. P. (1996). Seriously considering play: Designing interactive learning environments based on the blending of microworlds, simulations, and games. *Educational Technology Research & Development, 44*(2), 43-58. Retrieved August 7, 2005, from http://it.coe.uga.edu/~lrieber/play.html

Salen, K., & Zimmerman, E. (2003). *Rules of play: Game design fundamentals.* Cambridge, MA: MIT Press.

Schroeder, R. (1996). *Possible worlds: The social dynamic of virtual reality technology.* London: Westview Press.

Slater, M. (1999). Measuring presence: A response to the Witmer and Singer presence questionnaire. *Presence: Teleoperators and Virtual Environments, 8*(5), 560-565. Retrieved August 7, 2005, from http://www.cs.ucl.ac.uk/staff/m.slater/Papers/pq.pdf

Squire, K. (2002). Cultural framing of computer/video games. *Game Studies, 2*(1). Retrieved August 8, 2005, from http://www.gamestudies.org/0102/squire/

Squire, K. D., Makinster, J., Barnett, M., Barab, A. L., & Barab, S. A. (2003). Designed curriculum and local culture: Acknowledging the primacy of classroom culture. *Science Education, 87*, 1-22.

Tuan, Y. (1998). *Escapism.* Baltimore: John Hopkins University Press.

Weckström, N. (2004). *Finding "reality" in virtual environments.* Unpublished master's thesis, Arcada Polytechnic: Helsingfors / Esbo, Finland.

Wong, D. (2003). *The heritage & legacy of thinking and computer games. Online course essay.* Retrieved August 8, 2005, from http://www.msu.edu/user/buchan56/coursework/cep911_intellectual_history/hl_thinking.htm

Chapter VI

Inquisitivism:
The Evolution of a Constructivist Approach for Web-Based Instruction

Dwayne Harapnuik, University of Alberta, Canada

Abstract

This chapter introduces inquisitivism as an approach for designing and delivering Web-based instruction that shares many of the same principles of minimalism and other constructivist approaches. Inquisitivism is unique in that its two primary or first principles are the removal of fear and the stimulation of an inquisitive nature. The approach evolved during the design and delivery of an online full-credit university course. The results of a quasiexperimental design-based study revealed that online students in the inquisitivism-based course scored significantly higher on their final project scores, showed no significant difference in their satisfaction with their learning experiences from their face-to-face (F2F) counterparts, and had a reduction in fear or anxiety toward technology. Finally, the results revealed that there was no significant difference in final project scores across the personality types tested. The author hopes that inquisitivism will provide a foundation for creating effective constructivist-based online learning environments.

Introduction

The purpose of this chapter is to support my claim that inquisitivism (my adaptation of minimalism) is an effective constructivist online learning approach for adult learners who are required to learn new information technologies in a Web-based setting. Inquisitivism has emerged from the author's 10 years of experiences in course development and teaching in online and distance learning environments. Since the fall of 1996, over 3,600 University of Alberta students have completed either the full-credit undergraduate online course EDIT 435 or its graduate equivalent EDIT 535. These courses have been, and are still currently, delivered exclusively online with no F2F interaction.

They are officially called The Internet: Communicating, Accessing and Providing Information (Montgomerie & Harapnuik, 1996, 1997), but are colloquially referred to as "Nethowto," which is also the Web name of the course, and subsequently, the nickname that was adopted by students and faculty. In addition, several other courses based on the inquisitive approach have been designed and delivered by the author in both the academic and professional training environment.

This presentation of inquisitivism, its development, its application, and evaluation findings presented here are not based on a single case study or a "one-off," but are based on a body of data and experiences collected over a 10-year period. The inquisitivist approach was first formalized in 1998 (Harapnuik), was updated in 2004 (Harapnuik), and has been continually revised. Inquisitivism, and its application, continues to evolve in response to the needs of the author's primary academic responsibility — his students.

Constructivist Approaches Like Minimalism are Effective Foundations for Designing Technology Instruction

There is a body of literature that calls for a change in the way we design and deliver educational material: Objectivism vs. constructivism: Do we need a new paradigm? (Jonassen, 1991), Web-based distance learning and teaching: Revolutionary invention or reaction to necessity (Rominiszowki, 1997), The Learning revolution (Dryden & Vos, 1994), Transforming learning with technology: Beyond modernism and post-modernism or whoever controls the technology creates the reality (Jonassen, 2000), and Beyond reckoning: Research priorities for redirecting American higher education (Gumport, Cappelli, Massey, Nettles, Peterson, Shavelson, & Zemsky, 2002). The authors of these works argue that traditional forms of instruction are no longer effective. There are also claims that the deficiencies in the outcomes of learning are strongly influenced by underlying biases and assumptions in the design of instruction (Rand, Spiro, Feltovich, Jacobson, & Coulson, 1991). The systems approach to instructional design may be the primary factor contributing to the poor outcomes of instruction since it is still the predominant instructional design assumption used throughout most of education (Carroll, 1990; Dryden & Vos, 1994; Hobbs, 2002; Jonassen, 1997; Newman & Scurry, 2001; van der Meij & Carroll, 1995).

The systems approach is based on the assumption that learners are passive receptacles for information that the instructor (teacher or instructional media) relays. Educators are beginning to recognize:

...that our dominant paradigm mistakes a means for an end. It takes the means or method called 'instruction' or 'teaching' and makes it the end or purpose.... We now see that our mission is not instruction but rather that of producing learning with every student by whatever means work best. (Barr, & Tagg, 1995, p. 14)

Similarly, Carroll (1990) argued against the notion that learners are passive receptacles, and made a case against the systematic approach to learning, in his book the *Nurnberg Funnel*. The title refers to the legendary funnel of Nurnberg that was said to make people wise very quickly by simply pouring knowledge into them. The title is also a somewhat sarcastic accusation against traditional forms of instruction.

In the *Nurnberg Funnel,* Carroll presented minimalism as the culmination of 10 years of empirical research that showed that newer methods of instruction, based on constructivism and other cognitive theories or approaches, perform much better than the commonly used systems approach to instruction. Constructivists posit that knowledge is constructed, not transmitted, and that it results from activity. They also hold that knowledge is anchored in the context in which learning occurs, and that "meaning making" is in the mind of the knower, which necessitates multiple perspectives of the world (Jonassen, 1990, 1991, 1997). Meaning making is prompted by problems, questions, confusion, or even disagreement, and this meaning making is generally distributed or shared with others through our culture, tools and community (Jonassen, 1990, 1991, 1997; Kearsley, 1997; Strommen & Lincoln, 1992; Vygotsky, 1978).

Carroll's (1990) research revealed that instruction based on guided exploration (GE) was significantly more effective than the traditional systems approach. Out of a group of 12 participants at the IBM Watson research facility, 6 used (GE) cards, and the other 6 were given the traditional systems-style manual (SM). Both groups were expected to complete their respective training by working through either the drill or practice of the systems-style manual, or the 25 GE cards. Both groups were evaluated by being required to complete a real task of transcribing a one-page letter into a word processor and printing it out. The participants were asked to think aloud, and research associates recorded their thoughts. In addition, the sessions were videotaped so that all the data could be collated and taxonomized to develop a qualitative picture of how GE learning was contrasted by SM learning.

The use of guided exploration cards resulted in much faster initial learning, and more successful performance in the achievement task. The learning time for the GE participants, on average, was less than half of what it was for their SM counterparts: 3 hours and 55 minutes vs. 8 hours and 5 minutes (Carroll, 1990). Similarly, GE participants spent half as much time on the achievement task as did their SM counterparts, and the GE group achieved much greater success than the SM group. The GE group spent more time working on the actual system trying out more operations than the SM group, who spent most of their time reading about the system. Not only did the GE group work effectively with the operations needed to complete their task, they experimented with many more aspects of the system.

Carroll (1990) argued that the GE group was more successful because they worked with the system itself, and took responsibility for their own learning. They demonstrated much more initiative, and used errors as learning experiences. In contrast, the SM group often became trapped in error loops created by the systems-style manual. The problems the SM group experienced with the instructional material hindered or, in some cases, even prevented the learners from working with the system they were attempting to learn.

Carroll (1990, 1998) also argued that there is a need for a change in the way instruction is developed and delivered, and offered minimalism as a viable option for this change. An examination of the learning theory literature also reveals many theories and approaches to learning. A partial list includes structuralism, functionalism, connectionism, behaviorism, objectivism, and constructivism. When you add all the other theories that are not suffixed with an "ism" (classical conditioning, information processing model, etc.) there are over 50 learning theories and approaches (Kearsley, 1997).

Perhaps one reason that there are so many theories and approaches is that their authors have also sought out theories to substantiate or validate their research, and they, too, found that there was no single theory or approach that accurately supported or represented their work. When a suitable comprehensive theory or approach is not found, it is not uncommon for the researcher to propose new concepts, and combine elements of other theories and approaches into a new approach that could be applied specifically to a unique situation. This partially explains the creation of the inquisitivist approach.

Development and Evolution
of Inquisitivism

Inquisitivism is a descriptive approach to designing instruction. It shares many of the same principles as minimalism, but offers two key principles, or components, that set it apart. These two principles are codependent in the sense that the second principle cannot be realized without the first. The first principle of the inquisitivist approach is the removal of the fear that many adults have when first faced with learning to use technology. Many adults who are new to technology are virtually paralyzed when placed in front of a computer. The fear of "breaking something," or perhaps the fear of looking or feeling foolish, often prevents these adults from embracing computers and technology (DeLoughry, 1993; Shul & Weiner, 2000).

The second, most significant, or dependent, principle is the stimulation of inquisitivism. By designing instruction that reduces the "hurt level" and encourages the "HHHMMM??? What does this button do?" approach/attitude to learning, adults can be encouraged to learn in a similar fashion to how children learn (Harapnuik, 1998). Exploring and discovering the power and potential of computers, and technology in general, can be an exciting and stimulating process, if the learner is confident that they "can't break the system" or that the system "won't break them." With fear reduced and the inquisitive nature stimulated, it can be argued that adults can have almost the same level of success with technological learning as children. An inquisitivist approach to learning technology is essential because technology is dynamic and is rapidly changing, forcing learners to continually adapt to these changes.

Another significant factor about inquisitivism is that the approach was developed (and continues to evolve) during the development and continued delivery of the Nethowto Web-based course. The development of the inquisitivist approach was a practical response to a need, and was the result of a search for a theoretical foundation for the design, development, and delivery of the course. As *Nethowto* evolved, it became clear that many of the principles that ultimately became foundational to inquisitivism were at work in the development of the course.

In 1997 and 1998, the third and fourth year the Nethowto course was delivered, and the second and third year it was delivered exclusively online, the minimalist approach was researched, and even though it was originally designed as an approach for document design, components of its rubric seemed very appropriate to, and were applied to, Nethowto. During this time, it became apparent that even though minimalism satisfied many of the instructional design needs of Nethowto, and had the potential of providing a sound theoretical foundation for the course, it was lacking in two key areas: fear removal and social interaction. Kearsley affirmed the "solid theoretical foundation for minimalism" (Kearsley, 1998, p. 395), but also pointed out that it does have theoretical gaps. The most significant gap in minimalism is that it does not address the social aspect of learning (Kearsley, 1998). A lesser gap is that minimalism has not been tested in a variety of media, specifically online systems. As a result, the adaptation of minimalism proceeded, and inquisitivism was formalized in 1998 (Harapnuik, 1998). Table 1 offers a comparison of inquisitivism to the constructivist learning environments (CLE) and minimalist rubric from which it ultimately evolved.

It must be noted that many of the same principles apply to all three approaches. For example, all three approaches share the need for students to work on real-world tasks in genuine settings. As would be expected of constructivist approaches, all three emphasize knowledge construction, whether it is called reasoning and improvising, or discovery learning. Since inquisitivism is an adaptation of minimalism, it shares even more of the same principles. Inquisitivism is continually evolving, but there are currently 10 key concepts/components that make up the approach.

Table 1. Comparison of constructivist learning environments, minimalism, and inquisitivism

Constructivist Learning Environments	Minimalism	Inquisitivism
Provide multiple representation of reality	Reasoning and improvising	Fear removal
Avoid oversimplification of instruction by representing the natural complexity of the real world	Getting started fast	Stimulation of inquisitiveness
	Training on real tasks	Getting started fast
Present authentic task (contextualizing rather than abstracting)	Using the situation	Using the system to learn the system
	Reading in any order	Discovery learning
Foster reflective practice		
	Supporting error recognition and recovery	Modules can be completed in any order
Focus on knowledge construction, not reproduction		
	Developing optimal training designs	Supporting error recognition and recovery
Enable context-dependent and content-dependent knowledge construction		
	Exploiting prior knowledge	Developing optimal training designs
Support collaborative construction of knowledge through social negotiations, not competition among learners for recognition.		Forum for discussion and exploiting prior knowledge
		Real world assignments

Application of Inquisitivism to Nethowto

Carroll (1990) stated that taking checklists seriously is perhaps the most typical and debilitating design fallacy. Despite this strong statement, Carroll provided a rubric of minimalist principles. Similarly, inquisitivism has evolved into an approach with a rubric of principles. An early version of the following 10 principles was applied to the Nethowto course during a significant redesign of the course in the fall of 1998. It must also be noted that the course is still running, and both course and the 10 principles have continued to evolve.

Fear Removal

Dealing with the paralyzing fear that many adult learners experience must precede the stimulation of one's natural inquisitiveness. Demonstrating that the computer or any other piece of technology is not fragile, providing explanations, examples, and solutions for common errors and problems, and the application of data backup will help quell the adult learner's fear.

In an asynchronous education and Web-based environment, an instructor is not able to interact directly in person with an entire class (i.e., some students may be working in a different time zone) and to reassure the group as a whole. Nor can an instructor gauge body language, or tone and inflection of voice, to detect that fear may be an issue. Furthermore, both e-mail and Web-based conferencing interactions, which are essential to Web-based learning, are not direct forms of interaction, but are considered mediated transactions (Harasim, 1993; Lapadat, 2002). Because of these dynamics, fear or anxiety removal is perhaps one of the most challenging components to effectively facilitate, primarily because the F2F cues are missing, and students cannot be led through their anxieties. Using video or audio files to present what would be presented in a traditional F2F setting was, until recently, not a feasible option. While it is possible to use compressed video or audio to communicate with students now, there still is the issue of getting students over the initial fear or anxiety that they may have to operate this type of software for the very first time.

Because of these limitations, the asynchronous nature of the course, and the need to keep pages small to load quickly, the actual design and layout of the course's main Webpage had to be a primary factor in calming the fearful student. The main page (and the entire site for that matter), by design, is very simple and uncluttered. Students are not overwhelmed by choices on the main page, and a large "Getting Started" heading was strategically placed to be one of the first items noticed on the page.

The actual Getting Started instructions (referred to as First Steps) were broken down into four simple steps. The items in the four steps were designed to lead a student through the initial familiarization with the course. Students were not required to actually complete any assignments, but were still required to familiarize themselves with the course navigation and layout, to fill out a consent form (data was also used to create student profiles in the course administration system), to join the course conferencing system and, finally, review the introduction module.

The intention of the Getting Started page was that by following the four steps, fearful students would gain enough experience and success with the course to help them overcome or, at minimum, deal with their fear. While these four steps appear to be a linear systematic instruction (SI) type system superimposed on a minimalist structure, students can do the steps out of sequence, or ignore them all together, and still proceed through the course, so the sequencing aspect of SI is not a factor in student progression. At some point, and in some order, students will have to fill out the consent form, join the conferences, and begin work on the introduction module. These instructions are simply presented in their most logical order. Throughout the steps, students were encouraged to contact the instructor directly if help was needed. Students had (and currently still do have) access to the course instructor via e-mail, the Web-based conferencing system called the WebBoard (WebBoard Collaboration Server, 2005), and by phone.

Stimulation of Inquisitiveness

With the fear abated, the adult learner's intrinsic (but often suppressed) inquisitive nature can be stimulated and encouraged to flourish. Nethowto students are actually encouraged to read the "HHHMMM??? What does this button do?" approach article that is linked on the main page. The article details the 10 inquisitivist principles, and makes an argument for this approach as the basis for Web-based instruction.

The design of the course forces the students to make many more decisions, and to extensively investigate and use computer programs more than they are often used to. For example, in the first formal assignment, students are asked to submit an e-mail attachment, but they are not required to use a specific e-mail client or word processor. Students are directed to resources that they can use to learn about e-mail, e-mail clients, and the sending of attachments. In addition, students are required to investigate one aspect of attaching documents that most people take for granted, the encoding format. The only way that students can be sure that they submit an attachment in the required MIME encoding format is to explore the online Help within their e-mail clients or on the Web. This starts the whole inquisitivist process. Students quickly learn that a small amount of investigation within the programs they are currently using will reveal the results that they need. The immediate success students experience is a crucial aspect of inquisitivist design that will be further expounded in the getting started fast category.

Using the System to Learn the System

All training must take place on the actual system that is being learned. Every aspect of Nethowto is conducted online. Students are actually using the Internet, while learning about all forms of Internet communication, and accessing and sharing of information. In addition to the students conducting all aspects of the course online, the instructor of the course (the author) does not maintain an office at the University of Alberta campus, but conducts all aspects of design, development, and delivery of the courses completely online. In essence, the instructor uses the system to teach the system.

Getting Started Fast

Adult learners often have other interests than learning a new system. The learning they undertake is normally done to complement their existing work. The "welcome to the system," prefaces, and other nonessential layers in an introduction, are often ineffective uses of the learner's valuable time.

The Getting Started/First Steps sections of the course are designed to give students confidence in their initial experience with the course. The simple procedures that students are asked to follow, like joining the course conferencing system and using an online form to submit their student information, contribute positively to their learning experience. Similarly, all the information that students are required to review in the Getting Started section of the course is intended to contribute immediately and positively to their learning experience, and ultimately, give the learner confidence in the system.

The first assignment, submitting an e-mail attachment, is relatively simple to complete, and is strategically placed and used to give students immediate success. Students usually make the e-mail submission immediately after moving through the Getting Started section. A concerted effort is made to insure that students receive an immediate reply, and have rapid confirmation of their success. Students who have difficulty with the assignment are quickly directed to the resources that they need to use to have success in the assignment. The goal of the instructor is to reply to students within 3 to 4 hours of their first assignment submission (if the assignment is submitted during regular business hours, the reply is often processed in a matter of minutes).

Discovery Learning

There is no single, correct method or procedure prescribed in the course. Allowing for self-directed reasoning and improvising, through the learning experience, requires the adult learner to take full responsibility for their learning.

Throughout all course modules and course work, students are given specific assignment requirements that specify what should be submitted or included in the portfolio. Nethowto students are also given the freedom to choose the programs they use to complete the assignments. Unlike many technology-related courses that provide step-by-step instructions on conducting a specific procedure with or within an application, students are pointed to Web-based resources that deal more with the general concept than with the specifics of a particular application. This is not to say that step-by-step instructions are not necessary. There is a section of each module that points to links for the more common applications used in the course (FTP, Telnet, Text or HTML editors, etc.) that do provide the step-by-steps instructions for those who are most comfortable with this form of instruction, or are not comfortable with learning by doing, experimenting, or exploring.

All module coursework culminates in the course portfolio in which students have to display all they have learned in a Web site (part of the learning process is learning HTML). Students are told what is to be included in the portfolio, but are not explicitly instructed on how it should be created or formatted. Instead of a rigid recipe or formula, students are given the freedom to construct their portfolio in any way they choose. Links to instructional sites on HTML, Web design, graphics utilization, and usability are provided, but students are still required to learn the application of the technical aspects of creating a Web site to create their portfolios and projects. Marking guides, (details on what markers will be looking for) and examples of previous student work are provided to offer students additional guidance on what is ultimately expected. Although many students simply copy the format of previous student work, some students embrace this freedom, and come up with innovative ways to display their portfolios. These innovative portfolios are often included in the examples, but unfortunately, most students choose the safety of copying the simple or tried and true designs.

Modules can be Completed in any Order

Materials are designed to be read or completed in any order. Students impose their own hierarchy of knowledge, which is often born of necessity and bolstered by their previous experience. This helps to eliminate the common problems that arise from material read or completed out of sequence.

Providing a structure for openness requires a great deal of planning and structure. The course is modular, and each module, except for the portfolio, which is a compilation of all other modules, can be completed in any order. The module naming conventions do not include numbers or alphabets, to prevent any suggestion of a specific order. Despite the effort to not prescribe an order, and even though the modules can be completed in any order, most students follow the sequential listing of assignments in the course navigation structure. This, too, is part of the design. This order has been established for those students who lack confidence or experience with technology. By following the sequence of modules, students who lack technology confidence and experience can gain enough confidence and experience from the modules to successfully complete the portfolio and final project. While this sequential ordering of the modules may appear to be a linear SI type system superimposed on a minimalist structure, students can still do the modules out of order, so the sequential ordering of the modules is not as significant as it would be in a true SI system. Due to the very divergent capabilities of students in the course, the structure of the course has to serve both students with little experience, and those who may be very experienced. Students who need the order and structure can use the implied order from the navigational listing, and students who have the confidence to work on course modules in their own order have the freedom and opportunity to do so as well.

It must be acknowledged that even though there is no required order for completing the modules, the portfolio does require that the other minor assignment modules be completed first. A hierarchy of knowledge for the course is imposed by the two main course assignments. In order to complete the portfolio, students must learn HTML (hypertext mark up language) and complete the other assignments. In order to complete the final projects and earn a satisfactory grade, gaining experience in HTML development (either with a text or HTML editor) through building the portfolio is the most logical path for students to follow.

Supporting Error Recognition and Recovery

Errors must be accepted as a natural part of the learning process. Since there is such a pervasiveness of errors in most learning, it is unrealistic to imagine that errors can be ignored. Error recognition and recovery strategies need to be implemented to enable learners to learn from their mistakes, instead of being trapped by them. The use of FAQ's, Help Forums and other help strategies should be implemented to deal with the errors and problems that arise.

Once again, the asynchronous nature of Nethowto necessitates that the course itself provides support for error recovery. The Help link is strategically placed

1/3 of the way down the page and in the center (which is the area of the screen where a users eyes will first fall). The Web-based conferencing system and the Help conferences are also readily available. An online FAQ, and multiple admonitions to ask for help, are placed strategically throughout the course.

In addition to the actual design, layout, and structure of the course, the students are given immediate feedback (usually within minutes or, at most, hours) on their first assignments, and also receive detailed feedback (complete with written explanations) as to what mistakes were made on their portfolios. Students are encouraged to learn from their mistakes in the portfolios, and apply what they have learned to the final project. Students are given the option of submitting their portfolios 3 weeks prior to the end of term to receive an evaluation that will help prevent them from making the same errors on their final project that they made on the portfolios, and to give them a better of understanding of what is expected in the creation of a Web site.

When the students contact the instructor for help, they are first directed to the location in the course pages where the answer may lie. If the students report that they had reviewed the support material and were still not able to find a solution to their problems, they are then directed to additional support material where the answer could be found. If the additional support materials were not adequate, the students are then directed to even more information to help them determine the answer on their own. It is extremely important for the instructor to judge the level of frustration students may be experiencing and, if necessary, give them a direct answer sooner than later.

To insure that students help needs are met, all students are regularly queried about the course Website, and asked for suggestions on making changes to the course that would save them from having to contact the instructor, or use the Help forums for assistance.

Forum for Discussions and Exploiting Prior Knowledge

Adult education dealing with technology is often conducted through alternative delivery methods. Distance education, Web-based instruction, and other alternative delivery methods can isolate students. Providing a conferencing system for the replacement of F2F interaction is a crucial component of any alternative delivery program. Most adult learners of technology are experts in other areas or domains. Understanding the learner's prior knowledge and motivation, and finding ways to utilize it is one of the keys to effective adult training. In addition, adult learners can share their expertise, or assist each other, and should be encouraged to use the conferencing system to facilitate social interaction.

The WebBoard conferencing system is an effective forum for enabling students to provide each other with assistance. To encourage students to assist each other (not an easy thing to do in a competitive academic environment where students strive to be at the top of departmental or faculty mandated marks distributions) students are assessed a Help participation mark based on the quantity and quality of their participation — this mark is worth 10% of their final grade. One of the most common responses to the Help forums is how useful and helpful they are. It is not uncommon for a number of students in each session to state: "I could not have made it through the course without the Help forums." In addition to help related issues, students are required to start a topic discussion on an area that they are particularly interested in. This topic discussion is also required, and contributes toward the student's Issues participation mark.

The WebBoard forums are an example of what Vygotsky (1978) coined as social learning. In his theory, he stresses that social interaction is a critical component of situated learning because learners become involved in a "community of practice," and adopt the beliefs and behaviors of that community. Experts (experienced individuals) within the community often share the beliefs and behaviors of the community unintentionally, or model the proper conduct through their behavior. Newcomers interact with the experts, and then they themselves move into the community to become experts. This process can be referred to as legitimate peripheral participation, and occurs unintentionally (Lave & Wenger, 1990).

Some students who admit (in the WebBoard forums) to being normally reserved, or who might not even participate in a F2F setting, are encouraged by the equality they find in the WebBoard environment, and embrace this component of the course. It is not uncommon for these students to log on daily, and to participate in most (if not all) discussions. Students who may be near completion of the course often provide encouragement to students who have joined the course late or have simply started late. This exchange of information and knowledge, and sense of community, is one of the most positive aspects of this course. It is not uncommon for some students to go out of their way while traveling, to find a computer to log on and continue to participate in their special virtual community.

Despite never meeting the students F2F, it was possible for me to get familiar with the students through monitoring their e-mail and Web-based conferencing communications. In one sense, it may be easier to get a better understanding of a student's personality and needs than in a F2F setting because of monitoring all their Web-based communications. This advantage over the F2F setting is offset by the disadvantage of not being able to read students' nonverbal expression, body language, and general reactions.

Real World Assignments

"Make-work" (purposeless) projects are often not an effective use of a student's valuable time. All assignments must have a real world application.

All Nethowto assignments are genuine "real world" tasks that almost any information professional that uses the Internet as a tool would do on a daily basis. The Internet offers much more than the just the Web or e-mail, and students are required to use a variety of the Internet tools (Listserv, Usenet, Telnet, FTP, IM, HTML and Search engines) to complete their assignments, which focus on the information that can be gathered, shared, or moved using the assortment of Internet tools, rather than focusing on the tools themselves. The goal of the course is to give students experience in communicating, accessing, and providing information on the Internet. The emphasis is on the information, and not the tools used to access or provide the information. Technology is put in its place, and is relegated to its rightful role as an information access tool.

Optimal Training Designs

Feedback facilities, like online surveys or e-mail, should be used to allow learners to immediately provide feedback on any aspect of a program. Problems with instructions, assignments, wording, or other problems, should be immediately addressed and corrected. Instructional models are not deductive or prescriptive theories: they are descriptive processes. The design process should involve the actual learner through empirical analysis, so that adjustments can be made to suit the learner's needs. "Develop the best pedagogy that you can. See how well you can do. Then analyze the nature of what you did that worked." (Bruner, 1960, p. 89)

The Nethowto course has evolved to its present state because of the students who have worked through the course and provided feedback. Student feedback is immediately acknowledged, and if a particular portion of an assignment instruction (or any portion of the course for that matter) requires modification to bring clarity, this is done immediately. If the same questions are asked repeatedly, the subject of those questions is addressed, and that aspect of the course is modified to provide less confusion and to improve clarity. When significant changes are made as a result of student's feedback, announcements are made on the course News and Announcements page, to insure that all students are made aware of the change. Designing and developing an effective learning environment is a dynamic process that requires immediate responses to problems that arise. Students are encouraged to fill out detailed, online evaluation forms that provide additional information for continued improvements.

Delivery of Nethowto

Because the inquisitivist approach was developed through the delivery of the Nethowto course, it could be argued that the inquisitivist approach is not only an effective approach for the design of Web-based instruction, but it is also an effective approach for the delivery of Web-based instruction.

Another factor in the delivery of Nethowto is that the instructor (the author) does not maintain an office on the University campus, but works at a distance, and uses the same Internet tools that the students are required to use. Because the system (the Internet) is not only being used by the learners to learn the system, but also by the instructor to teach the system, the students are not asked, or required, to do anything that is not practical, or that is simply not possible with the Internet. Leading or teaching by example is often one of the most effective ways to lead and to teach. When the students learn that their instructor not only "talks-the-talk" but also "walks-the-walk" and is sensitive to the genuine problems that arise with Web-based instruction and communication (in the case of the instructor, telecommuting) because the instructor uses the same system that they do, attitudes toward the course and this approach to learning tend to become quite positive.

Necessity often breeds ingenuity. The evolution of the inquisitivist approach is tied so closely to the design, development, and delivery of Nethowto that one could argue that the approach itself evolved out of necessity. The 10 components of the inquisitivist approach are evident in the design and delivery of Nethowto (some more so than others), and while some of the components may be applied more effectively than others, they all combine to provide an approach to Web-based instruction that is practical and effective for the students and the instructor.

Evaluation of Inquisitivism

The evaluation of inquisitivism involved two phases, and employed both quantitative and qualitative measures. In the first phase, a quasiexperimental design (nonequivalent groups design) method was used to compare the grades of the final projects produced by a sample of Nethowto and comparison group students, and a comparison of the scores of the level of student satisfaction collected from both groups. The mark on the final project was used as a measure of student success in learning the concepts taught in the course, and ultimately, as a measure of the effectiveness of the instructional approach. Both the Nethowto sample and the comparison group involved undergraduate students enrolled in courses that had very similar content. Both the Nethowto and comparison group courses were designed to increase student Internet experience, knowledge, and communications skills.

To determine if students in the inquisitivist based Nethowto course had a reduction in fear of technology, students from both the groups were asked to complete three questionnaires: computer anxiety rating scales (CARS), computer thoughts survey (CTS), and general attitudes toward computers scale (GATCS) prior to the start of the course, and once again upon completion (Rosen, Sears, & Weil, 1987; Rosen & Weil, 1992).

The Nethowto sample differed from the comparison group in that they were required to take their course, while the Nethowto group chose to take the course as an elective. A second difference was that 45% of the comparison group students had taken one or two computer courses, and the rest of the comparison group had even more formal computer training (one student had a computer certificate). In contrast, 55% of the Nethowto group had no formal computer training, and the remaining students who did have formal computer training had taken only one or two courses. In addition, the comparison group was slightly younger (29 vs. 33), had a higher number of single students, with an even lesser degree of dependence (children). Another difference noted was that over half of the comparison group did not work, and the remaining portion only worked part-time. In contrast, over two thirds of the Nethowto group worked either full- or part-time. Finally, the Nethowto class was taught in conjunction with a graduate level class, which resulted in undergraduate and graduate student interaction.

The second phase of the evaluation included a student satisfaction analysis that was conducted over multiple sections of the Nethowto course, over a span of 4 years. This phase of the study also involved using the Keirsey Temperament Sorter (similar to the Myers-Briggs Type Inventory) to determine for what personality type inquisitivism is more appropriate. Both aspects of this secondary evaluation were only applied to Nethowto students.

Academic Success Comparisons

To compare the results of the final project scores for the Nethowto and the comparison group, Web sites, submitted by students from both groups, were evaluated on the same criteria. The mark on the final project was used as a measure of student success in learning the concepts taught in the course, and ultimately, as a measure of the effectiveness of the inquisitivist approach. Evaluators, who were "blind" to the group membership, used the same evaluation

Table 2. Final project scores for the Nethowto and comparison groups

	Nethowto (n = 54)	Control (n = 23)
Mean	37.27	28.96
Std. Deviation	4.69	4.32
Std. Error Mean	.64	.90
	t-test	
t	7.18	
df*	75	
Sig. (2 tailed)**	.003	
Mean Difference	8.21	
SE Difference	1.14	

*Equal variances
**p < .05

criteria given to students in both the Nethowto and comparison group, and scored the Web sites. The final project Web sites were scored out of 50 points, which was based on an assessment of the project's purpose, relevance, appearance, navigation, organization, level of difficulty, and content. Students were allowed to choose their own topics for the final project, to insure that motivation for the projects was high. One of the goals of the final project assignment was to demonstrate that the students could take all their newly acquired Internet skills and apply what they had learned in the course through the construction of a Web site. Assuming that this goal was met, and that students did demonstrate what they had learned in the course, the mean score of 37 (74%) on the final projects for Nethowto students demonstrated that these students had learned the course content, and were able to demonstrate their newly acquired abilities in the final project.

The first research hypothesis was whether students who learned the same course content via the Nethowto course would do better on the final project as those students who learned in a F2F model. The null hypothesis is rejected because an independent t-test (Table 2) revealed that there is a statistically significant difference between the mean final project scores for the Nethowto (M=37.27, SD=4.70) and comparison group course (M=28.96, SD=4.32), with the Nethowto students scoring higher.

Student Satisfaction Comparisons

To assess the level of satisfaction with their learning experience between the two groups, the means of the response to "Overall, this was an excellent course"

Table 3. Course satisfaction scores

	Nethowto (n = 54)	Control (n = 23)
Mean	4.24	4.13
Std. Deviation	.82	.81
Std. Error Mean	.11	.17
	t-test	
t	.54	
df*	75	
Sig. (2 tailed)**	.59	
Mean Difference	.11	
SE Difference	.20	

*Equal variances
**p < .05

were compared. Students in both the Nethowto and comparison group were given an universal student ratings of instruction (USRI) evaluation (University of Alberta Computer Network Services, 2004) form that included eight questions, near the end of the course, to assess the instruction they had received, and to assess how satisfied they were with their learning experience. The very short instrument (eight questions), the fact that students were still actively working on the course, and the comparison group's instructor having his students fill out the questionnaire during class time resulted in a high response rates for both the Nethowto and comparison groups.

Both groups indicated that they agreed that this was an excellent course: Nethowto student's average response to the question was 4.24, and the comparison group student's average response to the same question was 4.13.

An independent t-test (Table 3) demonstrates that there is no statistically significant difference between the mean final project scores for the Nethowto (M=4.24, SD=0.82) and comparison group course (M=4.13, SD=0.81), and we, therefore, fail to reject the null hypothesis. The lack of significant difference indicated that even though the Nethowto group satisfaction scores were slightly higher, the difference was not significant enough to argue that the Nethowto group was more satisfied with their learning experience.

Expanded Student Satisfaction Results

In addition to comparing the sample and comparison group results, the results of student evaluations of Nethowto undergraduate students, in multiple sections of the course, spanning a 4-year period, were examined. This supplement has been included to provide a broader perspective on the student satisfaction levels of Nethowto students over an extended period of time. It was also made possible because of the data collection instruments established when the course was originally set up, and that were unaltered in order to collect longitudinal data for future research. The Nethowto course remained fundamentally the same in terms of design, content, and delivery over this 4-year period. The changes or improvements made in the course during this time dealt primarily with issues of content clarity, and also reflected responses to changes in updates in software applications and systems.

Slightly more than 36% of Nethowto undergraduate students, from multiple sections of Nethowto, filled out a postcourse questionnaire over a 4-year period, resulting in a sample size of 258 for this analysis, resulting in an n of 258 for this analysis. The following six responses (Table 4) were selected and analyzed from the questionnaire because these questions dealt specifically with aspects of student satisfaction. More specifically, the questions dealt with student perceptions on the amount they learned in the course, how satisfied they were with the inquisitivist approach, and if they found the approach effective.

The course satisfaction was measured using a Likert scale (the Likert technique measures attitudes in which subjects are asked to express agreement or disagreement on a five-point scale) with one being the lowest level (strongly disagree), and five the highest (strongly agree). While the students found they learned a lot in the Nethowto course, they were not as positive with respect to the format and structure in which the course was delivered. Students agreed or strongly agreed that they learned a lot, would be willing to take similar courses online, and perhaps most importantly, agreed that the course helped them to significantly grow in their knowledge of computers and Internet, but they did not agree that the structure was conducive to learning. In addition, a SD of 1.17 on

Table 4. Student responses to questions about their satisfaction

Student response	Mean	SD	n
I learned a lot in this course	4.34	.87	258
I found the structure of the course conducive to learning.	3.85	.99	258
I would take other courses offered in this online, individualized instruction manner.	4.05	1.04	258
This course helped me grow from one level of knowledge about and familiarity with computers and the Internet to a significantly higher level.	4.36	.79	258
I found the learning theory (inquisitivism) used in this course to be effective for this type of instruction.	3.90	.91	258
I would have preferred to take this course via a traditional "lecture/laboratory" mode.	2.28	1.17	258

a mean of 2.28 indicated that even though, on average, the student responses were close to neutral or leaned slightly toward disagreeing that they would have preferred to take the course via a traditional lecture/lab format, there was still a significant proportion of students that would have preferred to take the course via a traditional lecture/lab format. This observation is similar to the results of Goodwin, Miller, and Cheetham (1991) and Lake (2001). Their research confirmed that students subjected to active learning instruction would have preferred the more traditional lecture format, despite having achieved greater success.

Reduction of Fear

The original study design included an analysis of the comparison and Nethowto groups, but because only 4 of the 23 comparison group students who completed the pretest surveys completed the posttest surveys, a comparison between the comparison group and the Nethowto was not possible. While the response rate from the Nethowto course was higher, only 11 out of 54 (20%) students completed the posttest anxiety surveys, and 10 of 54 completed the posttest thoughts and attitude surveys. The poor response rates of these posttest surveys negated any statistically useful data.

In response to this development, additional data were used to determine if there had been a change in anxiety or fear for Nethowto students as a result of the inquisitivist approach in a larger sample. Since the CARS, CTS and GATCS questionnaires, which were established when the course was originally set up, were left in place in order to collect longitudinal data, undergraduate Nethowto students from multiple sessions over a 4-year period were included in this analysis. Of the 479 undergraduate students who completed the Nethowto course during this expanded time frame, 162 students completed the posttest anxiety questionnaire, 168 completed the posttest thoughts questionnaire, and 170 students completed the posttest attitude questionnaire.

The increase in the response rate of 33% of the extended sample, compared to 20% in the original Nethowto sample, could be attributed to students being sent an additional reminder with their final project evaluations to complete the posttest questionnaires, and to an additional reminder being posted on the course conferencing system.

The anxiety levels are represented by a Likert scale with 1 (not at all) being the lowest level and 5 (very much) the highest. The attitudes toward computers are represented in a Likert scale, with 1 (strongly disagree) being the lowest level and 5 (strongly agree) the highest. The thoughts about using computer levels are represented by a Likert scale with 1 (not at all) being the lowest level and 5 (very much) the highest. Questions about thoughts and attitudes towards computers were included in two of the three surveys to help isolate the question regarding anxiety toward technology, and prevent any overlap in student responses.

Table 5 provides the mean scores for pretest and posttest attitudes and thoughts, which are virtually identical, while there is a difference between the pretest and posttest anxiety scores.

Table 6 provides ANOVA results. This analysis provides evidence of a statistically significant reduction in posttest anxiety scores ($p \leq .01$) in the expanded sample. A repeated dependent t-test would have yielded the same result as a repeated measures ANOVA of the means, and could have been used, but an ANOVA was used because it reduces the chance of multiple test error, and reduces Type 1 error. There was no significant difference in the pretest and posttest scores for attitude and thoughts toward technology.

While the hypothesis that students in the inquisitivist based Nethowto course had a reduction in fear of technology is supported in the expanded sample due to the anxiety findings, this result has to be viewed in the context of there being no significant difference in the level of fear of technology in the original sample group.

Table 5. Means scores and standard deviations associated with pretest and posttest anxiety, attitudes, and thoughts about computers

Test		Mean	SD	N
Anxiety	Pretest	1.76	.64	162
	Posttest	1.28	.57	162
Attitude	Pre-test	3.13	.39	170
	Post-test	3.14	.35	170
Thoughts	Pretest	2.83	.39	168
	Posttest	2.87	.37	168

Table 6. Sources of variance in pretest and posttest anxiety, attitudes and thoughts about computers

Variance Source	df	MS	F	p
Pretest vs. posttest anxiety	1	2.07	14.01	.004*
Within cells error	161	.15		
Pretest vs. posttest attitudes	1	3.43	.11	NS
Within cells error	169	.25		
Pretest vs. posttest thoughts	1	1.01	.13	NS
Within cells error	167	.22		

*$p < .05$

Table 7. Mean scores and standard deviations of personality types of Nethowto students

Personality type	n	Mean	SD
Guardian	40	35.03	3.548
Artisan	147	34.43	4.398
Idealist	53	34.15	5.379
Rational	133	34.52	4.403
Total	373	34.49	4.459

Personality Type Suitability

To determine if inquisitivism is appropriate for all personality types, Nethowto students from multiple sections of Nethowto were asked, at the beginning of the course, to complete the Keirsey Temperament Sorter (KTS) II (similar to the Myers-Briggs Type Inventory). Temperament type was used as a factor in an ANOVA.

Table 7 includes the Nethowto student final project mean scores, and the standard deviations for each personality type.

Notice the similarity of mean values in the personality types. While there were significantly more artisan (147) and rational (133) than idealist (53) and guardian (40) personality types, there is very little difference in the final project mean scores. An analysis of variance showed that no significant difference exists among the mean scores of the final project for the students with the four different personality types: $(F(3/369) = .303, p = .823)$, and have, therefore, failed to reject the null hypothesis.

These results indicate that since students from all four, personality types scored equally on the final project, the inquisitivist approach would be suitable for all four, personality types tested. Or, more specifically, the inquisitivist based Nethowto course may enable students from the four, personality types to score well in their assignments.

Not only did this study show that the online students did better on their final projects than the F2F students, it also showed that there are was no significant difference in the levels of learning experience satisfaction between the online students and the students in the traditional F2F classroom. It has also been shown that there was a reduction in student anxiety, and the achievement with the inquisitivist approach did not differ (in terms of final project performance) for the four personality types measured by the Keirsey Temperament Sorter.

Nethowto Students
Exceeded Expectations

The significantly higher final project scores from the online (Nethowto) students can be corroborated by a recent meta-analysis of distance learning research (Allen, Bourhis, Burrell, & Mabry, 2002; Allen, Mabry, Mattery, Bourhis, Titsworth, & Burrel, 2004). The mean scores of the Nethowto students' final projects were 17% higher than the comparison group. This difference is especially surprising given the fact that, on average, the comparison group students had taken more computer courses, and had less work and personal responsibilities.

The difference in scores between the Nethowto and comparison groups could have been attributed to a variety of factors. It may be the case that the Nethowto students motivation to do well in the course was higher because the Nethowto group chose the course as an elective, while the comparison group was required to take their course. Another factor affecting motivation could be related to the fact the Nethowto group was more mature, had greater marital and family responsibility, and could have been more accustomed to project work and independent learning.

Perhaps one of the most significant factors is time on task, which is a factor often not effectively controlled in quasiexperimental designs of educational research (Joy & Garcia, 2000). By its very design, inquisitivist instruction requires students to use the system while they learn the system. This translates into the Nethowto students spending virtually all their time on the actual task of learning to communicate, access, and provide information on the Internet.

In contrast, the comparison group students had traditional lectures, which meant that even though they could have been listening to Internet related topics, or even discussing these topics, they were not actually working on tasks relevant to learning how to use the Internet. Similarly, the time spent in labs for the comparison group also may not have been considered to be productive time on task, due to the systematic design of the comparison group course. With this design, students worked through lab assignments that followed the traditional step-by-step format. While this type of recipe learning does allow students to successfully complete assignments, it may not effectively foster knowledge acquisition, as minimalism would suggest.

This situation has been evident in the delivery of Nethowto. Some education students, who come into the Nethowto course having completed a prerequisite course that uses the traditional systematic approach, often have problems transferring or applying their experiences from the previous course to almost identical assignments in Nethowto. The only difference in the assignments is that Nethowto assignments do not follow the systematic recipe, and they allow the student to choose the program they should use to complete the assignment. While it must be acknowledged that this data is anecdotal, the incidents where this situation has happened have occurred enough times to warrant reporting and consideration for further investigation.

Another contributing factor that may explain the higher success of Nethowto students is that there could be significantly more direct instructor-student interaction. Direct interactions with the Nethowto instructor either fall into the category of e-mail, Web-based messages replies, or telephone conversations. Since Nethowto is conducted completely online, tracking the e-mail and Web-based conferencing interactions is very simple. On average, Nethowto students have 31 direct interactions with their instructor per session (academic term). The direct responses to student questions in the Web-based conferencing system have the advantage of being available and accessible for all other students to view at any time. Unfortunately, instructor involvement or interaction was not tested in the study, but one can assume that the number of direct interactions were much higher in the online course than they were in the F2F course.

Yet another possible success factor for the Nethowto students that was not controlled or tested was the collaborative aspect of the inquisitivist approach. Nethowto students were required to participate in a Help forum, and 10% of their final mark was also derived from this participation. Another 10% of their final mark was derived from the Issues conference participation, where students were required to start and moderate an issue of their choosing, and were required to participate in issues discussions with other students. In total, 20% of Nethowto students' final marks were from Web-based conferencing participation, so motivation to participate was quite high. While this was not controlled for and not tested, it may be speculated that the help and issues participation contributed significantly to the Nethowto students' acquisition of knowledge and final project success. Vygotsky (1978), and similar social constructivist theorists, stress the significance of social learning, and the transfer of knowledge and expertise through social interactions; therefore, it can be speculated that this dynamic applied.

A final contributing factor to the Nethowto students' success could be their involvement with graduate students in the conferencing component of the course. Since the undergraduate and graduate Nethowto students participated in the same conferencing forum, it may be the case that the graduate students attitude toward learning could have positively affected the undergraduate students.

While the author would like to posit that the inquisitivist approach was primarily responsible for the Nethowto student success, the aforementioned speculated factors need to be tested in further research. Regardless of the reason for their actual success, Nethowto students appeared to have learned the course material, and appeared to be satisfied with their learning experience.

Nethowto and F2F Students Learning Experience Satisfaction

Evidence showed that there was no significant difference in the learning experience satisfaction between Nethowto students and the comparison group students. The differences between the Nethowto and comparison group satisfaction mean scores were slight, with the mean scores for the Nethowto group being slightly, but not statistically significantly higher. In addition to students being satisfied, it can be shown that Nethowto students believed that they learned a lot, and that their knowledge grew significantly. The evidence from the supplemental questionnaire given to the Nethowto students suggests that the students not only learned a lot, they agreed that the course helped them to grow from one level of knowledge and familiarity with computers and the Internet to a significantly higher level.

The only question that did not have a clearly positive response was the question of whether or not students would have preferred to take the course via a traditional lecture/laboratory mode. Even though, on average, the student responses were close to neutral, or leaned toward disagreeing that they would have preferred to take the course via a traditional lecture/lab format, there was still a significant proportion of students that agreed, and would have preferred to take the course via a traditional lecture/lab format. Similarly, the average student response, which was slightly more positive than neutral toward the online format, the wide spread, indicated by a large standard deviation (1.17), suggests that significant numbers of students would have preferred the traditional format. The slightly positive leaning toward the online format may be accounted for by the fact that approximately half the students in the course were true-distance students, and had no choice in the format of their instruction, or were accustomed to the online format. In contrast, approximately half the students in the course were nondistance students accustomed to attending traditional classes on campus. The students who indicated a preference toward the traditional lecture/lab format, may have done so because they were accustomed to this form of instruction, or they simply found traditional instruction easier, and were more comfortable following a recipe. It may also just be the case that students simply do not like active learning. These factors could be taken into account in further research.

Overcoming Inquisitivist Approach Challenges

Even though the data reveals that students in the Nethowto course performed very well in their final projects, were as satisfied with their instruction as the comparison group, and it appears the inquisitivism is suitable for the four measured personality types, there are still challenges to the approach. For example, one of the most interesting paradoxical situations is that too many questions are asked by students who have simply not even read any of the instructions, and at the same time, not enough questions are asked students who are looking for the hidden challenge to the course. Another paradox involves encouraging student participation in the course conferencing system, while at the same time limiting excessive participation and competition. One of the most perplexing challenges is addressing the unique instructional needs of the vast diversity of students who take the course. Rather than view these issues as obstacles, these issues should be, and are, viewed as opportunities to make improvements in the design and delivery of Nethowto. Addressing these challenges, and many other challenges that have arisen in the development and delivery of Nethowto, will be addressed in future publications.

Further Research and Conclusion

Since the inquisitivist approach is new and an adaptation of minimalism, it could be argued that studies need to be run again (perhaps numerous times), but with much greater controls. Future investigations into the effectiveness of the inquisitivist approach would have to

- Employ true random sampling and statistically meaningful samples.
- Control for prior knowledge, ability, learning style, teacher effects, time-on-task, instructional method, and media familiarity.
- Use a comparison group for all aspects (i.e., personality).
- Use instruments with a sufficient number of items to increase reliability.
- Establish reliability scores on final projects.
- Consider using continuous data rather than discontinuous (i.e., use personality scores rather than 4-point scales).

However, even if these independent variables could be effectively controlled, their application would be artificial, calling to question the whole media comparison (Joy & Garcia, 2000).

Future research could also investigate the role of time-on-task, the impact of instructor-student and student-student interactions, and the effect of graduate and undergraduate student interactions. The affect of the instructor's personality and teaching style on the implementation and delivery of the Nethowto model could also be investigated. An even more perplexing area of future research would deal with the question of why students who demonstrated a high level of success and satisfaction with the inquisitivist approach would still have preferred a traditional form of instruction. Carroll found a similar phenomenon in his research that revealed that despite the success with minimalist documentation, people still claimed to prefer the traditional documentation (Carroll, 1990). Goodwin et al. (1991), and Lake (2001) also found that despite demonstrable improvement in achievement levels over lecture-based instruction, most students perceived active learning instruction to be ineffective, and would have preferred lecture-based instruction.

Are these claimed preferences actual preferences, or simply people's natural tendency or desire to preserve the status quo? Alternatively, does the inquisitivist approach and similar active-learning approaches expect, or require, too much of the learner? Are classes easier in the traditional systematic design format? Are inquisitivism, minimalism, active learning and many other student-centered constructivist approaches really such hard work, or are students simply more comfortable with memorization than with learning how to think? These questions are just the beginning of many more questions that would need to be effectively explored to determine why people appear to still prefer systematic design instruction, despite demonstrable success with other instructional approaches like inquisitivism.

Inquisitivism, minimalism, and active learning can be hard work, especially for those who are not accustomed to this form of instruction. Similarly, memorization is much easier than learning how to think critically and analytically if one is accustomed to memorization. We clearly need to change student's experience and perceptions towards these forms of instruction. Lake (2001) suggested that we expand the discussion for the rational of active-learning methods, incrementally introduce active learning and, finally, change to an all-active learning curriculum. I agree with Lake, but would add that we need to move toward a much broader adoption of inquisitivist, minimalist, and other forms of constructivist approaches at the primary and secondary levels, so that when students reach the postsecondary level, they are accustomed to the challenges and benefits of these active and engaging forms of instruction.

References

Allen, M., Bourhis, J., Burrell, N., & Mabry E. (2002). Comparing student satisfaction with distance education to traditional classrooms in higher education: A meta-analysis. *American Journal of Distance Education, 16*(2), 83-97.

Allen, M., Mabry, E., Mattery, M., Bourhis, J., Titsworth, S., & Burrel, N. (2004). Evaluating the effectiveness of distance learning: A comparison using meta-analysis. *Journal of Communication, 54*(3), 402-420.

Barr, R. B., & Tagg, J. (1995). From teaching to learning: A new paradigm for undergraduate, *Change*, (November/December), 13-25.

Bruner, J. S. (1960). *The process of education.* Cambridge, MA: Harvard University Press.

Carroll, J. M. (1990). *The Nurnberg Funnel: Designing minimalist instruction for practical computer skill.* Cambridge, MA: MIT Press.

Carroll, J. M. (1998). Reconstructing minimalism. In J. M. Carroll (Ed.), *Minimalism beyond the Nurnberg Funnel* (pp. 1-18). Cambridge, MA: MIT Press.

Carroll, J. M., & van der Meij, H. (1998). Ten misconceptions about minimalism. In J. M. Carroll (Ed.), *Minimalism beyond the Nurnberg Funnel* (pp. 55-90). Cambridge, MA: MIT Press.

DeLoughry, T. (1993). Two researchers say "technophobia" may affect millions of students. *Chronicle of Higher Education, 39*(34), 25-26.

Dryden, G., & Vos, J. (1994). *The learning revolution.* Rolling Hills Estates, CA: Jalmar Press.

Goodwin, L., Miller, J. E., & Cheetham, A. D. (1991). Teaching freshman to think: Does active learning work? *Bioscience, 41*(10), 719-722.

Gumport, P. J., Cappelli, P., Massey, W. F., Nettles, M. T., Peterson, M. W., Shavelson, R. J. et al. (2002). *Beyond reckoning: Research priorities for redirecting American higher education.* Retrieved March 18, 2005, from http://www.stanford.edu/group/ncpi/documents/pdfs/beyond_dead_reckoning.pdf.

Harapnuik, D. K. (1998). *Inquisitivism or "the HHHMMM??? What does this button do?" approach to learning: The synthesis of cognitive theories into a novel approach to adult education.* Unpublished manuscript, University of Alberta. Retrieved March 16, 2005, from http://www.quasar.ualberta.ca/edit435/theory/inquisitivism.htm.

Harapnuik, D. K. (2004). *Development and evaluation of inquisitivism as a foundational approach for Web-based instruction.* Doctoral Thesis, University of Alberta, CA.

Harasim, L. (1993). Collaborating in cyberspace: Using computer conferences as a group learning environment. *Interactive Learning Environments, 3,* 119-130.

Hobbs, D. L. (2002). A constructivist approach to Web course design: A review of the literature. *International Journal on E-Learning, 1*(2), 60-65.

Jonassen, D. H. (1990). *Computers in the classroom: Mindtools for critical thinking.* Englewood Cliffs, NJ: Prentice Hall.

Jonassen, D. H. (1991). Objectivism vs. constructivism: Do we need a new philosophical paradigm? *Educational Technology: Research and Development, 39*(3), 5-14.

Jonassen, D. H. (1997). A model for designing constructivist learning environments. In Z. Halim, T. Ottoman, & Z. Razak (Eds), *International Conference on Computers in Education* (pp. 71-80). Kuching, Sarawak, Malaysia: University Malaysia Sarawak & Asia Pacific Chapter of Association for the Advancement of Computing in Education (ACCE).

Jonassen, D. H. (2000). Transforming learning with technology: Beyond modernism and post-modernism or Whoever controls the technology creates the reality. *Educational Technology, 40*(2), 21-25.

Joy, E. J., & Garcia, F. E. (2000). Measuring learning effectiveness: A new look at no-significant-difference findings. *Journal of Asynchronous Learning Networks, 4*(1), 33-39.

Kearsley, G. (1997). *Learning & instruction: The theory into practice (TIP) database.* Retrieved March 16, 2005, from http://www.gwu.edu/~tip/

Kearsley, G. (1998). Minimalism: An agenda for research and practice. In J. M. Carroll (Ed.), *Minimalism beyond the Nurnberg Funnel* (pp. 393-406). Cambridge, MA: MIT Press.

Lake, D. A. (2001). Student performance and perceptions of a lecture-based course compared with the same course utilizing group discussion. *Physical Therapy, 81*(3), 886-902.

Lapadat, J. C. (2002). Written interaction: A key component in online learning. *Journal of Computer-Mediated Communication.* Retrieved March 16, 2005, from http://jcmc.indiana.edu/vol7/issue4/.

Lave, J., & Wenger E. (1990). *Situated learning: Legitimate peripheral participation.* Cambridge, UK: Cambridge University Press.

Montgomerie, T. C., & Harapnuik, D. K. (1996) The Internet: Communicating, accessing, & providing information [University of Alberta course Web site]. Retrieved March 16, 2005, from http://www.quasar.ualberta.ca/nethowto/.

Montgomerie, T. C., & Harapnuik, D. K. (1997). Observations on Web-based course development and delivery. *International Journal of Educational Telecommunications, 3*(2), 181-203.

Newman, F., & Scurry J. (2001). Online technology pushes pedagogy to the forefront. *The Chronicle of Higher Education, 5*, 7-11.

Rand, J., Spiro, R. J., Feltovich, M., Jacobson L., & Coulson, R. L. (1991, May). *Cognitive flexibility, constructivism, and hypertext: Random access instruction for advanced knowledge acquisition in ill-structured domains.* Retrieved March 16, 2005, from http://www.ilt.columbia.edu/ilt/papers/Spiro.html.

Romiszowski, A. J. (1997). Web-based distance learning and teaching: Revolutionary invention or reaction to necessity? In B. Khan (Ed.), *Web-based instruction* (pp. 25-37). Englewood Cliffs, NJ: Educational Technology Publications.

Rosen, L. D., Sears, D. C., & Weil, M. M. (1987). Computerphobia. *Behavior Research Methods, Instruments, & Computers, 19*(2), 167-179.

Rosen, L. D., & Weil, M. M. (1992). *Measuring technophobia: A manual for the administration and scoring of the computer anxiety rating scale (Form C), the computer thoughts survey (Form C) and the general attitudes toward computers scale (Form C).* [Manual, Version 1.1]. Dominguez Hills, Carson: California State University.

Shull, P. J., & Weiner, M. D. (2000). *Thinking inside of the box: Retention of women in engineering.* ASEE/IEEE Frontiers in Education Conference. Kansas City, MO: IEEE Education Society. Retrieved March 16, 2005, from http://fie.engrng.pitt.edu/fie2000/papers/1242.pdf

Strommen, E. F., & Lincoln, B. (1992). *Constructivism, technology, and the future of classroom learning.* Retrieved March 14, 2005, from http://www.ilt.columbia.edu/publications/papers/construct.html

University of Alberta Computer Network Services. (2004). *Universal ratings of instruction.* Retrieved March 1, 2005, from http://www.ualberta.ca/CNS/TSQS/USRI.html

van der Meij, H., & Carroll, J. M. (1995). Principles and heuristics for designing minimalist instruction. *Technical Communications*, *42*(2), 243-261.

Vygotsky, L. S. (1978). *Mind in society.* Cambridge, MA: Harvard University Press.

WebBoard Collaboration Server (2005). [Computer Software]. Calsbad CA: Akiva Corporation.

Chapter VII

Situated Learning and Interacting with/ Through Technologies:
Enhancing Research and Design

Pirkko Raudaskoski, Aalborg University, Denmark

Abstract

There is a growing interest within social and humanistic sciences towards understanding practice both theoretically and analytically. Lave and Wenger's (1991) concept, "situated learning," describes the process of newcomers moving toward full participation in a community. Wenger later refined his approach in his book Communities of practice: Learning, meaning and identity. *Situated learning is equalled with social order: instead of understanding learning as a separate practice from everyday life, learning is seen as a more mundane phenomenon. It is sometimes difficult to operationalize Lave and Wenger's concepts in data analysis. Ethnomethodology and conversation analysis (CA) find that social order is created continuously by its members in their interactions. As ethnomethodology and CA base their findings on rigorous data analysis, they are extremely useful in analysing situated learning in everyday practices. The interdisciplinary interaction analysis (IA) is suggested as the best way to study the various aspects of situated learning in technology-intensive interactions.*

Part 1: Theories of Situated Learning and Interpretation

The editors of this book, in their call for papers, were concerned with the gap that there seems to be with the theoretical and policy-oriented idea(l)s about technology and learning, and in the various educational establishments that have gone for the digital arena, and they were looking for contributions from scholars and practitioners that would discuss the problems. Frederick Erickson, in his recent book on talk and social theory, makes the important observation that innovation and change are not top-down processes, but that possibilities for societal change are deeply related to people's everyday experiences and conversations (Erickson, 2004). Thus, change and innovation are also bottom-up processes, though hard to capture or see if the analytical focus is not zoomed correctly. This paper addresses both methodological and practical (design) concerns: How can we best capture "situated learning" as an every (work)day experience, and how could this research practice help build better technologies.

With the increase in educational and other technology use, there is a growing need to do solid empirical research to uncover existing practices and design technologies that can support interaction and learning. This is why, in the following pages, I go through four approaches to social order and/or learning: Lave and Wenger's "situated learning," Wenger's "communities of practice," ethnomethodology/conversation analysis (CA), and interaction analysis (IA). I start with Lave and Wenger (1991), whose book on situated learning has been very influential with educators and theoreticians who have tried to capture the essence of learning in everyday life. The message I take from Lave and Wenger is that we should be aware of learning in circumstances other than in educational institutions. Then I discuss Wenger's later (1998) ideas about practice and learning. Next, I go through some basic ideas about conversation analysis and ethnomethodology, and discuss the empirical nature of their investigations. I end up with interaction analysis, which I see as useful if we wish to synergise the theoretical ideas of the mentioned approaches of social order and situated learning with concrete analytical tools. This way, the theoretical ideas and concepts of situated learning (and communities of praxis) become observable entities in concrete talk and action. Section two finishes this chapter with an example of a data analysis of an authentic technology-shaped and mediated interaction situation.

The aim with this chapter is to discuss Lave and Wenger's influential theories on situated learning, and in so doing, to be part of the "practice turn" in contemporary theory (Schatzki, Knorr-Cetina, & Savigny, 2001). There is a growing interest in linguistics, philosophy, sociology, and human sciences, in general, to better understand theoretically, for example, the relation between everyday practices, and social order and language use. However, it is crucial that with an interest in practices, the tools to analyse them are as fine-tuned as are the theories. Therefore, my chapter also contributes to the wider international research in which ethnomethodology and CA are used as approaches on computer-supported cooperative learning (CSCL) (cp. Koschmann, 1996), and computer-supported cooperative work (CSCW) (cp. Luf et al., 2003). In demonstrating how the basic ideas of "situated learning" and "communities of practice" find not only theoretical, but also analytical counterparts in CA/ ethnomethodology, my contribution also falls naturally between CSCL and CSCW.

Situated Learning and Social Order

All of the approaches to be introduced share the theoretical assumption that society is "done" in people's everyday practices. Lave and Wenger (1991) emphasise that not only is society produced and reproduced in schools, work-places, and homes, for instance, but that learning is an essential part of these activities. Therefore, it should also be natural that empirical studies are needed in order to better understand what these processes entail.

Situated Learning

Lave and Wenger (1991) regard situated learning as something that takes place in all the realms of human life. They stress the importance of the master-apprentice relationship in any learning situation. The newcomer will learn through participation in the community's everyday practices; in other words, by interacting with other people and human-made tools.

Traditionally, learning is considered as an individual, often cognitive, ability. However, Lave and Wenger (1991) change the analytic focus "from the individual as a learner to learning as participation in the social world, and from the concept of cognitive process to the more-encompassing view of social practice" (p. 43). In the following, the core concepts and ideas of Lave and Wenger's approach are explained.

Social Practice

Social practice is the site at which an individual meets others in a meaningful activity:

Briefly, a theory of social practice emphasizes the relational interdependency of agent and world, activity, meaning, cognition, learning, and knowing. It emphasizes the inherently socially negotiated character of meaning and the interested, concerned character of the thought and action of persons-in-activity. This view also claims that learning, thinking, and knowing are relations among people in activity in, with, and arising from the socially and culturally structured world. This world is socially constituted; objective forms and systems of activity, on the one hand, and agents' subjective and intersubjective understandings of them, on the other, mutually constitute both the world and its experienced forms. (Lave & Wenger, 1991, pp. 50-51)

Understanding and practice could be regarded as key concepts for learning: when we continuously make the world intelligible to ourselves and to others, through expressing our understandings of what is going on and how we understand the other people's understandings, we are—together with others, that is, socially — creating the situation as something, establishing social order. The understandings can change in brief moments or over time, and we act upon those understandings, which are visible in and as practices.

Participation

Lave and Wenger compare participation to meaning negotiation:

Participation is always based on situated negotiation and renegotiation of meaning in the world. This implies that understanding and experience are in constant interaction — indeed, as mutually constitutive. The notion of participation thus dissolves dichotomies between cerebral and embodied activity, between contemplation and involvement, between abstraction and experience: person, actions, and the world are implicated in all thought, speech, knowing, and learning. (Lave & Wenger, 1991, pp. 51-52)

Participation therefore fixes meanings for the purposes of the moment; meanings (not just of language, but of any object or situation) are never stable as such.

Legitimacy of Participation and the Nature of Knowledge

One of the basic concepts of Lave and Wenger's research is legitimate peripheral participation, or LPP, which accentuates learning as "an integral and inseparable aspect of social practice" (Lave & Wenger, 1991, p. 31). Instead of being taught about a community's ways of doing, a new member starts his or her "career" as an observer, who little by little is given more demanding tasks in the organisation. Language can be used to transfer knowledge, but explanations are bound to the local accomplishment of tasks, rather than being abstracted away as textbooks, and so forth. Lave and Wenger stress that as important as knowledge transmission, is learning how to talk, how to be a legitimate participant at the social interaction level.

Traditional teaching and learning relies heavily on abstractions or general knowledge. However, Lave and Wenger point out that knowing a general rule [which we could call also knowledge *that* (Ryle, 1975)] does not mean that we are able to apply it (cf. Ryle knowledge *how*), or even recognise it as relevant in specific circumstances:

In this sense, any "power of abstraction" is thoroughly situated, in the lives of persons and in the culture that makes it possible. On the other hand, the world carries its own structure so that specificity always implies generality (and in this sense generality is not to be assimilated to abstractness). (Lave & Wenger, 1991, p. 34)

Elsewhere, Lave and Wenger emphasise the situated nature of knowing, and call it an "activity by specific people in specific circumstances" (Lave & Wenger, 1991, p. 52). An example on many occasions is when knowledge, whether general or specified, is conveyed through instructions and examples.

Talking Within and Talking About

From the previous paragraph, we can summarise that all work and learning is based on participation, an integral part of which often is talk, and it is possible to differentiate between different types of talking. One distinction can be made between *talking within* and *talking about*:

Talking within itself includes both talking within (e.g., exchanging information necessary to the progress of ongoing activities) and talking about (e.g., stories, community lore). Inside the shared practice, both forms of talk fulfil specific functions: engaging, focusing, and shifting attention, bringing about coordination, etc. (Lave & Wenger, 1991, p. 109)

Here, Lave and Wenger come closer to what people are actually doing or saying in their everyday practice. That is, if we had a tape recording of people's talk, we could fairly easily pinpoint where these types of talk occur, and after some further analysis, to see what the function of the talk is at that point.

Summary

Situated learning takes place in everyday practices, and can be analysed as taking place at different levels, from momentary understandings to changes in practice. It is clear that both people's fleeting interactions and more stable formations, such as written documents and artefacts, are important for the unfolding of practices and understandings of them. However, Lave and Wenger do not go deeper into the division. Wenger (1998) takes the discussion a bit further.

Communities of Practice

Wenger (1998) aims at capturing, through the concept of community of practice, among other things, the essence of learning in communities, both from the aspect of the organisation (change) and the individual (identity). He takes, again, social practice as the essential site of meaning negotiation in a community. In brief, "practice" means "doing in a historical and social context that gives structure and meaning to what we do" (Wenger, 1998, p. 47).

Participation and Reification

"Doing" always implies action or active participation in the world, and therefore, it emphasises the participant's ability and task to make sense of the world indexically and continuously. However, this does not mean that we are "making things up as we go along," but we resort to existing human-made focus points such as language-in-interaction, written documents, sings, and so forth, which structure our world (thus, social agents and structure are in constant exchange). These are products of "reification." According to Wenger, participation and reification are always present in human activity: the continuous negotiation of meaning relies on these two phenomena. How do we, then, detect those in empirical data? Wenger does not want to be too abstract nor too detailed in his analysis of community of practice, which "is neither a specific, narrowly defined activity or interaction nor a broadly defined aggregate that is abstractly historical and social" (Wenger, 1998, pp. 124-125). So, Wenger claims, in this book, that individual conversations are too small entities to find evidence for learning, and organisational entities, such as nation or corporation, too crude categorisations to find out where learning really takes place. As Wenger's main interest seems to be how organisational practices change over time, interviews — in addition to observing the practices of the community — are regarded as a working level of analysis. However, a counterclaim might be in place: even if organisational practices and members' identities change over time, it should be possible to find traces or hints of that in even the smallest of exchanges. Also as important should be to gain a good understanding of how practices are reproduced, when there is no change — or is reproduction possible without any modification?

Identity as Practice

Wenger sees identity as fluid as any other meaning negotiation in practice:

Each act of participation and reification, from the most public to the most private, reflects the mutual constitution between individuals and collectives. (Wenger, 1998, p. 146)

Members of society have experiences that help them orient to, and understand, ongoing situations. Therefore, Wenger sees identity as:

a layering of events of participation and reification by which our experience and its social interpretation inform each other. (Wenger, 1998, p. 151)

This layering of events could be compared with Peircean semiotics and the endless interpretant: the thought/interpretation will continue changing according to previous thoughts/interpretations/signs. I have compared this individualistic, mentalist picture of interpretants with the publicly available, social turn-taking, and sequential interpretation: each turn is an interpretant that is built further by other participants in following turns (Raudaskoski, 2000a).

Thus, if participation and reification is going on in every occasion of talk (because language is a resource that has certain ascribed, even if negotiable and situatedly negotiated, meanings), then the potential for change and learning are also there. We just need good enough methods to be able to detect those phenomena that constitute towards change or learning in interactions. In other words, we need an analytically rigorous tool that is based on the idea that social order is "done" in interactions: ethnomethodology/conversation analysis, and the approach that draws from them and other interactionist approaches, with a special emphasis on the embodied and material nature of interactions: interaction analysis.

Ethnomethodology/Conversation Analysis

Ethnomethodology grew as a challenge to traditional society, which saw and sees social order or structure as something that is decisive for what people do. Harold Garfinkel and Harvey Sacks are the founders of ethnomethodology, and the ensuing conversation analysis (CA) that took talk as the primary target of investigation. What is studied is the continuous production of social order in people's everyday lives. According to these domains, we orient to the world as intelligible and predictable, and at the same time, produce the world as understandable for us, and others, through our actions and interactions. This "documentary method of interpretation" (Garfinkel, 1967) takes place all the time, for instance, in human-computer interaction. One of the early successes in humans interacting with computers was Weizenbaum's ELIZA, a "Rogerian therapist" that would engage in discussion with a user[1]. The success of the programme, which is based on simple pattern matching, can be attributed to the fact that users would orient to whatever was said as a "therapist" speaking, and therefore, would see the sometimes odd repetitions or phrasings as "therapeutic talk" rather than as out of place. Thus, the "pattern" of therapeutic interactions that the people had an idea of beforehand would explain the "features" that they encountered in each response from ELIZA.

This basic idea of an amalgamation of the actors' context/world, or their interpretation of the world, compares with what Lave and Wenger have said about participation: experience and interpretation are in continuous exchange. So, the documentary method of interpretation offers a good basic explanation of what Lave and Wenger's social practices are.

Sequential Interpretation as Realised Through Turn Taking

The documentary method of interpretation is realised in sequential interpretation: our actions show how we understand the situation, and form a basis for what the next interpretation will be about. In talk, this means that each turn is an interpretation of the previous one, and is again interpreted in the ensuing talk by others.

The emphasis in CA/ethnomethodology on the indexical nature of meanings that are negotiated in sequential interpretation resembles Wenger's idea about language as a reification (basic meanings), the context-bound meanings of which are negotiated through participation. But indexicality of meaning is not limited to individual words or sentences: also such linguistic activities as stories, descriptions, and explanations achieve certain meanings indexically. Something is "done" with them, in a specific interaction (see Lave & Wenger's *talking about* and the contextual nature of knowledge).

When tape recorders became commonly available in the 1960s, Sacks and his fellow researchers could record conversations and start analysing them as instances of social order. It is important to note that all the observations that CA claims about conversational interaction are based on data analysis, not on "armchair philosophising" (cp. speech act theory). They found out that in everyday conversations, turn taking has some general features: everybody has the right to talk and talk as much as they want; turns are built up from so-called turn constructional units (TCUs), at the end of which a speaker change can occur (transition relevance place, TRP); current speaker can select the next one, and if this does not happen, people can self-select, otherwise the current speaker continues (Sacks, Schegloff, & Jefferson, 1978).

Turn taking is thus the technique through which people make sense out of each other, and of the activity in question. One way of talking about shared resources in interaction is "intersubjectivity":

Organizational features of ordinary conversation and other talk-in-interaction provide for the routine display of participants' understandings of one another's conduct and of the field of action, thereby building in a routine grounding for intersubjectivity. This same organization provides interactants the resources for recognizing breakdowns of intersubjectivity and for repairing them. (Schegloff, 1992, p. 1295)

The technology of turn-taking, with the help of which we speak and act, manages, at the same time, intersubjectivity, and shows that we can detect any problems in it. Through repair work, the otherwise unnoticed meaning negotiation becomes visible. Repair work has been divided into the following types, in the order of preference (on the basis of interactional work[2]): (1) self-initiated self-repair, (2) other-initiated self-repair (different conventions), and (3) other-repair (Schegloff et al., 1977). Thus we can (1) adjust our own talk, (2) tell others that there is a problem with their talk, and (3) modify others' interpretations.

Conversation analysis offers a reliable analytic tool to detect how the participants understand each other, and how they observe and repair any problems in conversation. Thus, this method can show how, in concrete interactions, a novice is made into a legitimate participant (i.e., a producer of social order), and also how, in repair sequences, a novice is shown how his or her understanding differs from that of the experts.

Charles Goodwin was one of the first researchers within the field to incorporate the visual aspect of data, the embodied nature of our interactions. This development co-occurred with the advent of video cameras. So, the interest turned to, for example, how speakers construct turns according to the actions of the hearers (Goodwin, 1979). Thus, the recipient design of a turn is potentially in constant revision. The idea that turns are only spoken material can also be amended: a nod constitutes a turn in the same way as does *yes* as an answer (Raudaskoski, 2003, discusses the impact of the visual on conversation analytic analyses of a computer tutorial use).

Membership Categorization Device

Membership categorization device (MCD) was a term introduced by Sacks in his lectures (Sacks, 1992). Membership categories are sets or social types that can be used to characterise an individual. So, when we talk about "a colleague," "Pirkko," "the Finn," or "that woman," we do not just refer to somebody (or, rather, select from a certain category set — see Sacks, 1992, Vol. 1: 41), but we also position the person in a certain way, for example, as an insider or outsider.

MCD is thus a feature that is observable within a turn; it is also recipient designed, as it could be possible that the speaker would choose different MCDs for a different audience. However, it is also possible to see how the MCD works in the sequence of turns: (how) does the next speaker orient to the specific formulation? MCD gives us a tool to analyse how identity is constructed, a topic of interest in research into technology-mediated interaction.

Summary

What can be learnt from ethnomethodological and conversation analytic research is that identity and change are ongoingly constructed in interactions; they are observable phenomena. This gives us a powerful tool to continue from the (still) overarching definition Lave and Wenger, together and separately, give to identity and meaning construction, and hence learning: they are continuously negotiated phenomena that constitute and change social practices. With ethnomethodology and CA, we can start looking at exactly how these phenomena look in our everyday action, and also what exactly can count as these phenomena. The latter means that not just are we able to analyse instances, the shape of which have accounted for *a priori*, for instance, identity construction, but we can also discover new, previously unidentified ways of negotiating the category of identity.

Interaction Analysis

Charles Goodwin was mentioned previously as an instigator of video analysis within CA. Another approach has been launched by Brigitte Jordan and Austin Henderson, who introduced, in their 1995 paper, interaction analysis (IA), a useful approach to researching people's actions in technological environments. They combine ethnography with other observational approaches, such as conversation analysis, to achieve reliable analyses of human interaction, learning, and technology use. In practice, this means that the researcher will want to know as much as possible about the research site and the members of the community, to have an understanding of the practice they are involved in. Thus, even though situated action is the locus of analysis, the participants' previous experience and routine ways of doing things are documented also outside of the situation under scrutiny, for instance, through ethnographic observation and interviews. This knowledge helps, then, in pinpointing the nature of situated learning, for example, in relation to a change in practice or identity construction. When written documents and other artefacts become an important part of analysis, it is good to know how those objects function, and came to exist in the organisation.

Analytic Foci

The present paper has not given many tools for doing data analysis. The reason is that if we acknowledge that meanings are negotiated locally, then most of the analytic schemas formulated for analysing discourse turn out not to acknowledge the situated nature of meaning and understanding. Only conversation analysis and ethnomethodology have taken the local meaning making seriously, in that they started with the conversational data, and not a model of it, trying to answer the questions: what are these people doing, how are they doing what they are doing, why are they doing this now? This approach is demanding, but it has started to reveal exactly how social order is done through interaction, and also how grammar and interaction constitute each other. Jordan and Henderson's analytic foci use the results and phenomena of interest from this work. These are features of interaction that can be, at least, the starting point for analysing interactions in technology environments.

The Structure of Events

By "event," Jordan and Henderson refer to an activity that has certain boundaries: for example, a meeting or advice-giving situation can be an event. Any event has to start and end, and often has some kind of segmentation within it. Participants achieve the beginning and ending of an event, and they transfer from one segment to another. It is important to find out how technology affects these accomplishments.

The Temporal Organization of Activity

There are two types of temporal organization that affect an activity. At the macro level, most of the members of the society have to take into consideration various schedules: meetings have been arranged to start at a certain hour; school lessons last for a certain length, and so forth. Here, the focus of analysis can be how the scheduling is visible, observable, has an effect on the ongoing interaction. However, temporality is an inherent phenomenon in all actions and interactions: talking and cooperating is done in a similar way (repetition), and there is a certain rhythm to it that an accomplished member has to manage. There are also periods of work followed by pauses in any work environment. The research question then becomes: how are these local temporal phenomena organised in technology-filled environments?

Turn Taking and Repair Work

Turn taking and repair were discussed already above. Often turn taking is understood to be only relevant for the "technique" of taking the floor or giving it. However, it is important to remember also what is being done through this technique: interpretations are continuously checked. A thorough understanding of turn taking is needed when any interactional phenomena is analyzed; for example, the phenomenon of "interruption" takes place only when a turn constructional unit (see previous) is intruded.

Jordan and Henderson include repair work in their list of analytic foci, as well: When there are any problems with the production or interpretation of turns, people can resort to repair work, which also shows how intersubjectivity is being managed in sequential, turn-by-turn, interpretation work.

Participation Structures

Participation structures are also called participation frameworks in conversation analytic literature. This concept refers to whom the participants are orienting in situations when there are more than two participants. Also, attention can be paid to artifacts, which then can become relevant in participation structures (cf. *talking within*). Participation structures can help understand what is going on in multiparty conversations or in workplaces with several people and technological devices: who or what is the recipient of talk or other action.

The Spatial Organization of Activity

When we interact with others in shared space, we organize our bodies and distances in certain ways. This choreography is also something that is negotiated: we change our posture, come closer, or shun away, according to what we regard as appropriate closeness. The use of space can also force certain social constellations: in a meeting, the chairperson sits at the end of the table, and therefore can see everybody and be seen by everybody.

Artifacts and Documents

Much attention in communication studies has focused on what people say. However, even in situations where talk is important — for example, meetings — documents and artifacts (cf. Wenger's reifications) are an important part of the ongoing action. A meeting agenda that everybody has in front of them is a case in point: the talk of the meeting is structured according to the list. The list can also be pointed at, or somehow else made relevant during the talk. This is what happens with artifacts as well: they are objects that are made part of an activity, either as the medium of the activity, or as a communicative resource (e.g., pointing at a computer during talking).

Summary

If the understanding of social practice and learning is such that it is ongoing — it is produced and reproduced in people's actions, but is not ahistorical — then interaction analysis offers a platform from which to research, also, those instances of human practice in which technology plays a crucial role. To be able to define what learning or change in practice could be, it is necessary, first, to find out how people actually interact through and with technology, and how they do that as embodied participants, not just language-using members. This is why videoed data of interactions are needed, and also detailed analyses of what is going on. When observations without video are done, not only are details of action missed, but also the phenomenon itself, and its categorization, might be problematic: what do the participants do that could then be categorized, for instance, as "impatience?" This might sound like hair splitting to people who want to understand the "big picture," but I would claim that that picture still emerges from the small pieces of a jigsaw puzzle, or should we say from the even smaller pixels that form the picture. If we do not understand what those are, that is, how the big picture or parts of it comes to be constructed during the action itself, then we might create a picture that has little to do with the empirically emerging one.

Part 2: A Case Study

Information technology is changing the landscape of practices in homes, workplaces, and educational institutions. The transformation has been fast, not only because the Information Society has become a goal on the agenda of many Western governments. This has led to a need to understand the impact of technologies on private and public life. There are many levels at which the opportunities and hindrances that technology brings with itself can be detected: from doing questionnaire research that reaches to thousands of people, to trying to understand the practice in a certain home or workplace. When the "grain size" is that of a participant's actions as they evolve (vs., for example, how he or she talks about them in an interview research), then video footage of the practice is the best way to capture what is going on. This is also true for finding out what situated learning and interpretation is in practice (pun intended).

Case: Collaborating Through Technology[3]

The object of the analysis is synchronous interaction within a group of collaborating scientists. Communication takes place through a desktop videoconferencing system with the following utilities: an audio connection, a low frame rate video connection displaying "talking heads" of the participants, and a presentation and interaction tool (referred to as the "whiteboard"), where participants can upload files with text and graphics, type, and draw, using keyboard and mouse respectively. I examine videoed sessions in which an "official" participant meets the others through the system, but also in which he has to deal with the people who are co-present in the room. This elaborate communicative situation with varying resources for participation and understanding offers an interesting scenario, a careful analysis of which can help us appreciate the complexity of the situated and embodied nature of the encounter.

The Data

The video data was not collected by myself, but was gathered for research purposes during an EU-funded project (1996-1997) MANICORAL (Multimedia and Network in Cooperative Research and Learning, see Nielsen, Duce, Knudsen, Sünkel, & Robinson, 1995). During the project, a group of scientists, dispersed in various parts of Europe, would meet through a computer system (MERCI). Some of these meetings were videotaped. I got access to two video recordings from a meeting in which a Danish PhD student, "Svend," is giving a presentation through the system, and who is accompanied by his supervisor and two other researchers who represented the human-computer interaction part of the project. One of the recordings came from the site where the talk was given, and the other from another Danish location, where a geologist was connected to the system.

The session lasted about 1.5 hours in all, and most of the time, the PhD student was active in giving his presentation, using the whiteboard for sharing slides. There were some technical difficulties, however, and it is during those episodes when both the people sharing the location with the PhD student and the distant participants would interact with each other. One of these "problem solving" occurrences lasted for 10 minutes, and was chosen to be analysed closer.

Data Analysis

A close interaction analytical scrutiny of the 10-minute extract revealed some interesting aspects of how the participants negotiated their, and others', participatory statuses through use of pronouns and body posture. Rather than regarding these phenomena as psychological, as reflecting Svend's intentions or strategies, the analysis aims to show the impact of the local interaction on what is said and done.

The extract is also interspersed with various types of instruction giving, an activity that is considered central in education and training, too. The analysis will show what kind of advice was given, and how the specific communicative circumstances shaped that activity.

The participants manage the communicative situation through talk, pointing, turning of heads, and body posture: the interaction is embodied, and each participant's actions are a communicative resource to the others to interpret what is going on. The same resource is available to the analyst, as is the other participants' next moves. It is from this texture of participants' interpretations that the analysis draws "red threads."

A fairly long extract from the 10-minute segment is discussed, as it demonstrates most of the issues that the analysis brought forward. The English translation of the Danish is placed in the transcript, such that it would not disturb reading what was said. Turns crucial for the analysis either have a box around them (the use of pronouns) or have a grey background (body posture)[4]. In the extract, Peter[5] and Volker are interacting from other locations through the MERCI system; Svend is a Danish PhD student; Jens, his supervisor; and Hanne, a member of the human-computer interaction research group. Figure 1 gives an idea of what the situation was like for Svend:

Who is the Participatory Status Oriented To?

Wenger's identity as participation/reification, and ethnomethodology's membership categorisation device accentuate the local, negotiated character of identity formation. Lave and Wenger discuss the newcomer's participation as legitimate and as peripheral, but legitimacy is also an issue in "central" participation. In the situation at hand, Svend is a ratified participant (together with Peter and Volker), a status that is achieved in the interaction. The extract also shows how technology, a material reificiation, shapes the local identity formation by giving Svend a possibility to fluctuate between being the ratified participant (*I*) or one among many (*we*). The participatory status is also done through bodily orientation, gestures, and gaze.

Constructing the Ratified Participants Through Language Use: We vs. I

As mentioned above, the room in which the primary data recording was done, was occupied not only by the PhD student giving a talk about his research, but also by his supervisor, and two of the HCI project members. In the other recording that we have for our inspection, the scientific researcher is also accompanied by other people, this time two HCI project research assistants who were doing the video recording. Though in that video recording, Peter, the

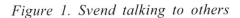

Figure 1. Svend talking to others

Extract 1.

```
1   Peter: i'll quit it and start it up with rat.

2   Svend: ((two head nods)) OK PETER ((sits back in chair=))

3   Jens: [=der kunne vi altså ikke høre/then we could not hear]

4          [((sits back in chair))]

5   V:     svend it's in options (.) under the rat,

6          [>in the< rat (.) erm:]

7   Svend: [((head nods))]

8   V:     windows [you have the option button.]

9   Jens:          [tsk= [han går meget fin igennem/he is very easy to hear]]

10  Svend:               [((=gaze to Jens and back to screen))]

11  V:     when you press rats there you can (.) select some uh

12         [redundancy algorithm,]

13  Svend: [((slight body and gaze turn toward Jens, hands into a 'ball'

14         shape))]

15  V:     [can you select this.]

16  Svend: [((eyes to screen, hands into a 'clap' position))]

17         (.)

18  Svend: ((moves to use the mouse; gaze to mouse,

19         [clicking mouse gaze to screen))]

20  Jens:  [((moving nearer the screen))] prøv i/try in opt[ions]

21  Svend:                                            [ER-] VOLKER, I

22         HEAR YOU VERY:: CL- L- CLEAR AND LOU^DLY NOW. ((hand away from mouse))

23         (2)

24  Jens:  ( åbne/open  )[heh ((outhaling))]

25                       [((gaze to Svend,))]

26  Svend:               [((turns to Jens))]
```

continued on following page

Extract 1. cont.

```
27
28              [hvad?/what? ((smiling))]
29      Jens:   [((gaze to screen))]
30
31              [(  åbne/open   )]
32      Svend:  [((gaze to screen))]
33      Peter:  can you hear me now?
34      Jens:   ja/yeah
35      Svend:  ((hand to mouse; leaning nearer to screen)) VI KAN GO-/WE CA- WE CAN
36              HEAR YOU PETER,
37              (.)
38      Jens:   ((whispering)) fordelen dag da kunne han ikke høre os
39      V:      did you understand, under the options button [(.)]
40                                      [((menu appears on the WB))]
41      S&J:                            [((gaze to menu))]
42      V:      you can select the redundancy,
43      Jens:   yeah (.)yeah (.) yeah
44      Svend:  YEAH,
45              (3)
46      Jens:   >which one.<
47      Svend:  AND WHICH ONE SHOULD WE: SELECT,
48              (2)
49      Hanne:  ( )((comes closer to screen to watch))
50              (2)
51              ((menu moved to the right of the WB))
52      Hanne:  ((goes away from screen))
53      Svend:  ((glances around the screen)) SORRY VOLKER BUT- ER WHICH (.) KIND OF
54              REDUNDANCY SHOULD WE SELECT,
55              (2)
56      V:      um the first one, i think.
57      Jens:   >p c m,<
58      Svend:  ER::: THAT WITH ER NONE
59      V:      no (.) no (.) the first one (.) that's not none so p c m.
60      Svend:  OKAY WE CHANGE TO P C M ((mouse clicks))
```

geophysicist, orients to the students, especially, by looking at them or grinning at them, and sometimes talking in Danish, which they can overhear: he always uses *I* when he participates in the discussion through the system. However, in the breakdown talk, Svend frequently uses *we* (which refers to the group in the room instead of the group of scientists online), in contrast to the use of *I* in his presentation. This usage highlights the nature of the breakdown discussion in which the whole group participates in the room (unlike what happened in Peter's room). Extract (1) shows the first exchange, in which *we* appeared in the 10-minute data extract.

The extract displays the impact of the two participation structures, or frameworks, that Svend oscillates between. When he talks to the distant members, not only does he look at the screen, but also his voice volume is strong (as indicated by small capitals), and when he interacts with the others in the room, the volume goes down, even to whispering sometimes. In line 24, Jens, Svend's PhD supervisor, says something after a 2-second pause in the interaction. He turns his gaze to Svend, thus changing his focus of attention from the object he is talking about to the recipient. Svend turns to Jens at the same time, acknowledging, thus, his status as the recipient of Jens talk. He initiates a repair sequence (*hvad*? what?), and orients to Jens' laughter by smiling. Svend's question means that he did not hear what Jens said, and thus, the smile indicates that he anticipates Jens' turn to be something potentially funny. When Svend's gaze reaches Jens, the latter immediately looks at the screen, and repeats what he said before. By turning his gaze to the screen, Jens accentuates the matter-of-factness of his comment (which seems to concern a menu on the screen), and cuts down any trajectory of an "aside," an entertaining remark about some aspect of the situation. Svend promptly follows Jens' gaze and turns to the screen again. So when Peter's question comes in through the system, Svend has been interacting with Jens, who answers to Peter's question in Danish. Jens' answering *ja,* "yeah" is appropriate in the sense that *you* in Peter's question could be plural and not singular, that is, it can be understood to refer to Svend or the whole group. Jens treats the situation to be something in which he is an active participant, not just an observer. However, he does not take over Svend's position as the primary recipient of Peter's talk, that is, he does not answer the question loud in English:[6] it is meant to be heard only locally. The intensity of the local interaction carries on to Svend's answer, which he, at first, starts formulating in Danish (Peter is Danish, too), and it also shows in his use of *vi* and *we*. The local participation framework affects Svend's changing his construction of the situation from himself (*I*) being the primary recipient to the whole group (*we*) being the target of the question.[7]

The use of *we* is maintained through a couple of more contributions from Svend (in lines 47, 53-54, and 60). The second one (*and which one should we select*) is a question addressed to Volker, who has repeated his instructions about selecting the redundancy level (and thus repaired a possible misunderstanding by Svend). Jens' *yeah, yeah, yeah* (line 43) displays his very eager formulation: he is urging Svend to give the answer. Jens continues his active role by prompting Svend's next turn (*which one*, line 46). Volker does not come immediately back with an answer, and the question with the *we* formulation is repeated (lines 53 and 54).

The last use of *we* in this exchange is in *okay we change to p c m*, which finishes off the exchange with the technical expert Volker. It is interesting that also this verbalisation of action (which in practice means selecting a certain option by a mouse click) is phrased as something done by the whole group, though it is only Svend that uses the mouse.

The technical fact that Svend has to do a mouse click before he can talk means that the onlookers have a good opportunity to produce the relevant next turn. In this exchange, Jens was a very active participant, and his "feeding the words" to Svend must have added to Svend's shaping his turns such that he is talking for the whole of the group, not just for himself. But it has to be noted that Svend was able to use the formulation also because in line 39, Volker's specific inquiry about Svend's understanding did not address Svend by name (*did you understand*), unlike at the beginning of the extract (line 5). To that appellation, Svend replies with first person pronoun (line 21). So it is not just what is going on in the room, but also the interaction between the distant participants that shapes how Svend orients to the situation. Jens gets more and more active during the breakdown, and towards the end of the segment, he is actually typing on the whiteboard, and even asks Peter if he can hear. But he is not the primary ratified participant in the session, which shows in his formulation of the question (which he repeats twice): *peter, can you hear now?* He does not use *me*, nor *us*, and thus, does not construct himself as the primary ratified participant ("I"), nor even as someone who can talk for the whole group ("we").

Participatory Status Through Body Orientation

The way the participants were sitting around the computer (see Figure 1) constructed Svend as the primary ratified participant in this technology-mediated discussion. And he was the performer of the meeting, as he was giving a scientific talk to the other members. However, as was already seen in the preceding analysis, the group in the room was very active, and Svend had two major participation frameworks to switch in between. The people in the room would not treat themselves as primary ratified recipients, as they did not talk directly to the distant members, but prompted Svend to say things. Only once did Svend offer Jens the keyboard to type to the whiteboard, and after that, Jens also talked through the system, as mentioned above.

Svend had three possibilities to bodily orient to what was going on in the room: (1) he could listen to and talk to the others while watching the screen; (2) he could indicate that he is orienting to the others in the room by tilting his head, or body, or both towards the others, while still gazing at the screen; or (3) he could turn to look at the colleagues in the room. The last option is the "strongest" in displaying orientation to the participation framework he is part of, in the room. In the data extract, the turns of interest, as regards Svend's body posture, are shaded grey.

The exchange in lines 26 to 32 was discussed in the previous analysis. It suffices here to say that difficulties in hearing what the other people in the room said, usually resulted in Svend turning to the speaker when he initiated repair.

Svend's head movement in line 10 (gaze to Jens and back to screen) exemplifies how he tries to manage being part of two competing participation frameworks: those of listening to Volker, and listening to Jens. Right after Jens' overlapping talk with Volker's in line 9, Svend turns his gaze to Jens, and orients back to the screen or Volker: Svend acknowledges Jens' turn while listening to Volker. In lines 13 and 14, Svend anticipates a possible end of Volker's turn (in conversation analytic terms, the end of a turn constructional unit), but is still listening to him: Svend indicates aligning to Jens by turning his body and gaze slightly toward Jens, and turning his hands into a position of, as if holding a ball. Svend's orientation is brought back to the screen by Volker's request in line 15, *can you select this*: he gazes at the screen, and the hands collapse together into a clap. As his next action is a mouse click, it looks like he interprets Volker's instruction in line 15 to be something to do immediately. However, it turns out that he clicks the microphone button to tell Volker how well his voice carries over.

Summary

The interaction analytical investigation of the data extract does not attempt to give definite answers to what types of communication take place in the new technology-mediated learning and working environments. Rather, it aims at showing how a detailed data analysis like this can help empirical researchers to identify phenomena, to see how the participants' practices really are situated, that they emerge moment-by-moment in the sometimes chaotic surroundings of using the technology. It is important to realise that the communication technologies do not provide a stable way of communicating: the participants' interactions are shaped by the resources offered by the technology, but also, and always, by the contingent interaction between the local and distant participants in the situation. Svend is the scientific and technological apprentice in the situation, and he has to balance between the participation frameworks with the distant members who include Volker, the technology expert, and the local members who include Jens, the PhD supervisor. He manages his physical and social position through embodied interaction. Svend's talk and head movements, together with his body orientation, make it possible for him, sometimes, to orient to both participation frameworks at the same time. The breakdown segment abounds with suggestions for what Svend should say to the distant members. This is made possible because of the technology: Svend has to do a mouse click before he can speak, which gives the others a time lag to come in with their suggestions. Thus, the cooperative nature of the turn construction is not a purely social phenomenon either, but afforded by the design of the system.

As mentioned previously, in the other recording, the participant always talked about *I*, and therefore the participatory status of the two research assistants remained that of an outsider or an observer. In Svend's case, the people in the room were sitting fairly close to him (so the physical distance was small), and they also were people he knew quite well (the social distance was small), whereas Peter was sitting a fair distance away from the students, and they did not know each other that well. Physical closeness in Svend's room meant that the "observers" could see what was happening on the screen, and their proximity to Svend made interacting with him natural. They also knew Svend, which made it easier for them to engage in what he was doing: they treated him as a representative of their community of practice, rather than the only legitimate participant in the meeting over distance.

Instructing in CMC

The analysis above shows that the participatory status was created through interaction, and by means that show its elusive nature. More "visible" issues that the breakdown segment concerned were: (1) the status of hearing and being heard, and (2) instructions of changing the audio settings of each terminal. Both topics are realised as questions and requests, actions that also take place in traditional educational settings.

We can now have a closer look at how the establishing of the participatory status, and these different types of actions, were related to each other. In practice, this means that the analysis deals with the same data, but this time, in order to concentrate on the nature of instruction giving.

Peter's question *can you hear me now?* (line 33) and the ensuing answer were a typical adjacency pair that served the purpose of checking the availability of one to the other distant participants. But in extract (1), there is another assurance about hearing that does not seem to fit in (line 21): it is given after a request to undertake an action, not about the status of hearing. For convenience, the beginning of extract (1) will be reproduced in extract (2):

Extract 2.

```
 1   Peter: i'll quit it and start it up with rat.
 2   Svend: ((two head nods)) OK PETER ((sits back in chair=))
 3   Jens: [=der kunne vi altså ikke høre/then we could not hear]
 4          [((sits back in chair))]
 5   V: svend it's in options (.) under the rat,
 6          [>in the< rat (.) erm:]
 7   Svend:   [((head nods))]
 8   V: windows [you have the option button.]
 9   Jens:      [tsk= [han går meget fin igennem/he is very easy to hear]]
10   Svend:          [((=gaze to Jens and back to screen))]
11   V: when you press rats there you can (.) select some uh
12          [redundancy algorithm,]
13   Svend: [((slight body and gaze turn toward Jens, hands into a 'ball'
14          shape))]
15   V:     [can you select this.]
16   Svend: [((eyes to screen, hands into a 'clap' position))]
17          (.)
18   Svend: ((moves to use the mouse; gaze to mouse,
19          [clicking mouse gaze to screen))]
20   Jens:  [((moving nearer the screen))] prøv i/try in opt[ions]
21   Svend:                                      [ER-] VOLKER, I
22          HEAR YOU VERY:: CL- L- CLEAR AND LOU^DLY NOW. ((hand away from mouse))
23          (2)
```

Volker starts his turn by addressing Svend by name (line 5), so the recipient of the turn should be clear. However, Svend seems to orient to Volker's turn as an informative, whereas Jens interprets it as action instigating ("try in options")[8]. The last word of Jens' turn (*options*), however, overlaps by Svend's *er*, that is, Svend has already started his turn to Volker and is not an active listener of Jens' instruction. Jens' instruction ("try in options") treats Svend's reaching for the mouse as an indication that Svend is going to do what Volker just told him to do. However, this is not the case, because Svend clicks the microphone button to comment about Volker's sound quality. Though Jens' turns in lines 24 and 31 are almost unhearable, in line 31, he seems to be repeating his turn from line 24, and the turn contains the word "open." So, most probably, Jens is commenting something about doing what Volker asked Svend to do (which would explain why Jens drops the trajectory of laughing, as it would be laughing at Svend, rather than with him).

Svend's orienting to the sound of Volker's turn is more understandable if we look at what Jens says just before and during Volker's turn: he is commenting about hearing and sound (lines 3 and 9). Thus, though Svend reacts to Volker's turn, he also belongs to the participation framework in the room, the topic of which had been the sound connection. This framework "overrode" that of Volker's, and therefore, his answer to Volker (starting line 21) is understandable in the room, but not for Volker. However, Jens orients to Svend's comment as out of place (in the context of the distant communication) by his outhaled *heh* (line 24), a sound that actually lies in between a laughter token ("hah") and a disagreeing "hmph."

Before the beginning of extract (1), Peter had been told to change to the RAT option. Thus, Volker's turn (lines 5 to 15) could be interpreted as a "presentation" about how to change to RAT (especially as Svend and Jens had just earlier come to the conclusion that they are using RAT, in other words, that is something they would not need to change to). If Volker's turn is heard as a general description rather than an instruction to do, then *can you select this* can be understood as a self repair by Volker: *you can (.) select some uh redundancy algorithm can you select this*. Volker did not have a falling intonation before *can you select this*, that is, his intonation did not indicate that *can you select this* would be a new utterance (and, also, *can* was said in lower volume).

So, Jens' active role in lines 43 (*yeah yeah yeah*) and 46 (*which one*) can be an exhibit that he wants to make sure this time Svend does not give "a wrong answer." Svend's *yeah* is not immediately followed by action, which indicates that he has answered a question about understanding, rather than interpreted Volker's turn as a request to do. Jens takes the next step by asking *which one*, showing that he is aware of the "doing" aspect of the instruction. It is interesting that Svend does not take up Jens' clear suggestion *p c m* (line 57), but gives his own interpretation about Volker's *the first one* (line 56). In the circumstances, Jens' *p c m* is ambiguous (as are all Jens' "answers" to the distant participants): (1) It can be heard as Jens' interpretation of Volker's *first one*, but, also, (2) as a suggestion for Svend's next line. By not repeating what Jens suggested, Svend does not integrate Jens turn to his primary participation framework, that is, Svend gives Jens' turn the first interpretation (i.e., what Svend does next is "legitimate" in the situation). Svend's own interpretation of *the first one* is what he sees as the first option in the menu, *none*, and this is what he puts forward. So, at a point where there was a clear difference of the distant participant's meaning, Svend would proceed with his own agenda. However, the common perspective is returned to with the *we* in *okay we change to p c m*.

Summary

During the session, Svend was given many kinds of instructions, both through the system and from the local participants. Instructions can be ambiguous: they can describe what to do, or they can be meant to be acted out immediately; the ensuing action sequence can thus be *assimilative* or *enactive* (Harris, 1996). Therefore, instructions can be related to *knowledge that*, to general information and knowledge, and *knowledge how*, the practice and skills of doing. This "theory" of explanations and instructions from the distant sites, especially from Volker, had to be made into "practice": a menu had to be opened, a mouse click to take place. Or, the listener indicated that he had understood the speaker[9]. But also more subtle instructive communication (or, rather, what was taken as instructive) was going on: Jens' "stories" about what had happened before. These remarks could be considered as *talking about:* however, they are closely connected to the topic of what was being done, and therefore become part of the talking within, or "exchanging information necessary to the progress of ongoing activities" (Lave and Wenger, 1991, p. 109). Svend interpreted these stories as being central to the ongoing communication, as was shown by his orienting to the status of the sound of the distant member, rather than to the instruction given. Thus, already a 1-minute extract could demonstrate that not only is power of abstraction thoroughly situated (Lave & Wenger, 1991, p. 34), but in how different activities knowledge transfer can happen.

Throughout the instruction giving, also interpersonal or social relations were at stake. This was reflected in Jens' withdrawal from a potentially embarrassing situation of laughing at his PhD student's mistake. This incidence also serves as an example of how Svend was treated as a legitimate participant: by not laughing at him, Svend's PhD supervisor did not engage in blame (c.f. Lave & Wenger, 1991, p. 111).

Discussion

In relation to learning, the primary ratified participant has the resource of the local group to aid his interaction with, and interpretation of, the distant participants: the local participants would volunteer information and next actions by feeding, in next turns, what Svend should say next, and by prompting him to check the audio connection. But they also were a "knowledge reserve" to Svend: he could turn to them (either physically or metaphorically) to negotiate what the distant participants meant. This constellation of the two interactions, in the room and through the technology, provided for an efficient and communication-bound way of dealing with problematic issues: situated negotiation was embedded within interaction with nonpresent participants. This is situated learning through, and at, new media.

Conclusion

Learning is taking place all the time, in educational and other institutions, and in everyday life. A careful investigation of how participants, in fact, "do" social practice, can thus be undertaken at the level of a fairly small "grain size." In this situation, technology played a decisive role. I hope to have shown how such concepts, from the situated learning/communities of practice approach as legitimacy of participation; talking within and talking about; and the nature of knowledge, reification, and identity as practice ("power"'), are realised in, and through, the sequential interpretation that takes place through turn taking. Awareness of this level can also be of help in designing for material reifications that best support dispersed communities of practice: However vast and global the influence of educational media, its use is rooted in everyday practices. Empirical research is also crucial for discussing and adjusting Lave's and Wenger's theoretical concepts. But, most importantly, empirical research, of the kind advocated in this paper, gives us a glimpse into the *how* of the practices in various technology-induced (learning) situations. Policy-making often is based only on the theoretical or "self-evident" knowledge *that*: there is a world out there and we know enough of it to push the changes forward. It is, however, advisable in the first place to understand what those practices we are trying to change actually look like, and how does technology then shape the practices, how do people manage *talking within* or *talking about*, instructing, and other mundane activities, when they are done through and at technology.

References

Coulter, J. (1994). Is contextualising necessarily interpretive? *Journal of Pragmatics, 21*(6), 689-698.

Erickson, F. (2004). *Talk and social theory: Ecologies of speaking and listening in everyday life.* Cambridge, UK: Polity Press.

Garfinkel, H. (1967). *Studies in ethnomethodology.* Englewood Cliffs: Prentice-Hall.

Georgsen, M., & Raudaskoski, P. (2002). What is local practice in technology-mediated environments? Video analysis of situated interaction and learning. In L. Dirckinck-Holmfeld & B. Fibiger (Eds.), *Learning in virtual environments* (pp. 309-330). Copenhagen: Samfundslitteratur.

Goodwin, C. (1979). The interactive construction of a sentence in natural conversation. In G. Psathas (Ed.), *Everyday language: Studies in ethnomethodology* (pp. 97-121). New York: Irvington.

Jordan, B., & Henderson, A. (1995). Interaction analysis: Foundations and practice. *The Journal of the Learning Sciences, 4*(1), 39-103.

Koschmann, T. (1996). Paradigm shifts and instructional technology. In T. Koschmann (Ed.). *CSCL: Theory and practice of an emerging paradigm* (pp. 1-23). Mahwah, NJ: Lawrence Erlbaum.

Lave, J., & Wenger, E. (1991). *Situated learning*. Cambridge, MA: Cambridge University Press.

Luff, P., Hindmarsh, J., Heath, C., & Hinds, P.C. (2003). Workplace studies: Recovering work practice and informing system design. *Computer Supported Cooperative Work, 12*(1), 123-125.

Nielsen, J., Duce, D., Knudsen, P., Sünkel, H., & Robinson, K. (1995). *MANICORAL, multimedia and network in cooperative research and learning, project contract*. EU: Telematics program. RE 1066.

Raudaskoski, P. (2000a). Between reading and encountering in human-(technical) text interaction. *Reader: Essays in reader-oriented theory, criticism, and pedagogy, 42*, 30-61.

Raudaskoski, P. (2000b). The use of communicative resources in videoconferencing. In L. Pemberton & S. Shurville (Eds.), *Words on the Web. Computer mediated communication* (pp. 44-51). Exeter: Intellect.

Raudaskoski, P. (2003). Users' interpretations at a computer tutorial. Detecting (causes) of misunderstandings. In C.L. Prevignano & P.J. Thibault (Eds.), *Discussing conversation analysis. The work of Emanuel A. Schegloff* (pp. 109-139). Amsterdam: Benjamins.

Ryle, G. (1975 (1949)). *The concept of mind*. London: Hutchinsons of London.

Sacks, H., & Jefferson, G. (Eds.). (1992). *Lectures on conversation* (vols. 1-2). Oxford: Basil Blackwell.

Sacks, H., Schegloff, E., & Jefferson, G. (1978). A simplest systematics for the organization of turn taking for conversation. In J. Schenkein (Ed.), *Studies in the organization of conversational interaction* (pp. 1-55). Volume I. New York: Academic Press.

Schatzki, T., Knorr-Cetina, K., & Savigny, E.V. (2001). *The practice turn in contemporary theory*. London: Routledge.

Schegloff, E. (1992). Repair after next turn: The last structurally provided defence of intersubjectivity in conversation. *American Journal of Sociology, 97*(5), 1295-1345.

Schegloff, E., Jefferson G., & Sacks H. (1977). The preference for self-correction in the organization of repair in conversation. *Language, 53*, 361- 82.

Weizenbaum, J. (1966). ELIZA — A computer program for the study of natural language communication between man and machine. *Communications of the ACM, 9*.

Wenger, E. (1998). *Communities of practice: Learning, meaning and identity*. Cambridge, UK: Cambridge University Press.

Endnotes

¹   It is possible to try ELIZA online, for instance, http://www-ai.ijs.si/eliza/eliza.html.

²   In repair, self-initiation is preferred to other-initiation: the space for both types of initiation are within three turns (the present, at the transition relevance place, and in the next turn) (Schegloff et al., 1977). In other-initiation, the trouble source can be made explicit, and there usually is a short pause, indicating a possibility for self-initiation, and thus making it visible that self-initiation (and correction) are preferred. Anything in the conversation can be a repairable.

³   In Georgsen and Raudaskoski (2002), the same situation is analysed from the point of view of Herbert Clark's theory of grounding.

⁴   The rest of the transcription conventions are as follows:

Name	distant participant
lower case (bold)	what was actually said
SMALL CAPS (BOLD)	talking to the distant members (louder voice)
x̲	stressed (part of) word
°word°	word delivered quieter than the surrounding talk
.word	word produced with an inbreath
>word<	speech item delivered quicker than other talk
.tch	"tisking" the tongue against the roof of the mouth
not in bold/NOT IN BOLD	translation of Danish into English
!	exclaiming tone of voice
.	falling intonation
,	flat intonation
(N)	length of pause in seconds
(.)	pause shorter than one second
=	talk/action latches on another
( )	analyst not sure what was said
(( ))	an activity or comment on the delivery of speech
[ ]	simultaneous speech/activity (in overlapping brackets)

5   The names are fictitious.

6   The fact that a mouse button has to be pressed when speaking through the system means that Jens could not choose to be a real, primary speaker (though it would be possible for him to indicate his willingness to be one by talking loudly).

7   The impact of local talk to a desktop videoconference interaction is also discussed in Raudaskoski (2000b).

8   Raudaskoski (2000a) and Raudaskoski (2003) discuss the interpretation of requests as instruction to read or instruction to do in two different language technology environments.

9   Coulter wants to keep a distinction between understanding and interpreting a text. According to him, understanding means that the reader is able to make the text intelligible, but interpreting is an activity in which the text is given a significance (Coulter, 1994). A parallel could be drawn between understanding and interpreting instructions here; the significance of an instruction would be acting it out, whereas understanding would refer to the more cognitive phenomenon of understanding what was said.

Chapter VIII

Rethinking E-Learning:
Shifting the Focus to Learning Activities

Jørgen Bang, University of Aarhus, Denmark

Christian Dalsgaard, University of Aarhus, Denmark

Abstract

"Technology alone does not deliver educational success. It only becomes valuable in education if learners and teachers can do something useful with it" (OECD, 2001, p. 24). This quotation could be used as a bon mot for this chapter. Our main goal is to rethink e-learning by shifting the focus of attention from learning resources (learning objects) to learning activities, which also implies a refocusing of the pedagogical discussion of the learning process. Firstly, we try to identify why e-learning has not been able to deliver the educational results as expected 5 years ago. Secondly, we discuss the relation between learning objectives, learning resources, and learning activities, in an attempt to develop a consistent, theoretical framework for learning as an active, collaborative process that bears social and cultural relevance to the student. Finally, we specify our concept of learning activities, and argue for the educational advantages of creating large learning resources that may be used for multiple learning activities.

Setting the Scene:
E-Learning Reconsidered

At the height of the dotcom bubble, around year 2000, many politicians and heads of educational institutions firmly believed that e-learning, online learning, virtual university education, or as sometimes called by real enthusiasts, "webucation," was the future solution to university education. In 1997, Peter Drucker, a business guru, predicted that "universities won't survive … as a residential institution" (The Guardian, 2004). Others, arguing along the same lines, foresaw that universities would become content providers and learning facilitators to for-profit producers of "learningware."

In the US, several universities formed commercial companies, alone, or in collaboration with other universities and cultural institutions. Among these were New York University (NYU) Online, and Fathom, formed by Columbia University, together with 14 universities, libraries, and museums: but they stopped before launching e-learning courses. One of the few successful e-learning providers in the U.S. is the University of Phoenix. Their success seems to be related to a focus on a limited and specialised market within IT, business, and health.

In March 2000, in Lisbon, the European Council adopted a grand scale plan named e-Europe — An information society for all, with the goal to become "the most competitive and dynamic knowledge-based economy in the world" by 2010 (European Commission, 2000). To achieve this, ICT and e-learning should play an essential role, and increase the participation in higher education to approximately 50% of a youth generation. In May the same year, the European Commission published a communication named eLearning: Designing Tomorrow's Education (European Commission, 2003).

Parallel to these political initiatives, but without coordination from the side of the Commission, many national projects for e-learning were launched in the first years of the century, for example, the UK e-University, the Digital University in The Netherlands, the Bavarian Virtual University, the Finish Virtual University, and the Net-University in Sweden.

In 2005, 5 years later, the UKeU has ceased operation. What the Minister of Education then launched, as a worldwide twenty-first century successor of the Open University, never attracted financial support from commercial partners, and only recruited 900 students, at a time when 5,000 were expected. A huge investment spent on developing a dedicated, new, learning platform, and a focus on the supply side (selling courses from partner institutions) instead of on student needs, seem to have been the basic reasons for the collapse. A market analysis was never performed.

Other European virtual university initiatives are facing similar problems, but not for the same reasons. The Dutch Digital University, a consortium of universities in The Netherlands, together with some IT-companies and publishers, has not reached a significant volume. The Finish Virtual University, a government initiative to enhance collaboration within the higher education sector in Finland, has increased the amount of online courses, but there has been very little extended collaboration between institutions as intended, and no considerable increase in intake of new student groups. Similar developments have occurred for the Bavarian Virtual University and the Swedish Net-University, also government initiatives with similar goals as the Finish Virtual University: to enhance institutional collaboration and increase the offer of e-learning opportunities. The allocated funding has primarily gone to development of online courses for full-time students. Thus, the flexibility seems to have increased, but not the interinstitutional collaboration, nor the recruitment of new student groups. None of these European e-learning initiatives have reached a level of sustainability, and they will not survive if government support is withdrawn.

Shifting the Focus from Technology to Pedagogy

In a report published as early as 2001, the OECD predicted this development:

In spite of having spent US$16 billion in 1999 in OECD countries on ICT, there is little evidence that ICT meets the original promise of better education for more people at less cost. As a result there are now concerns over the return investment.

(...) There is however no clear evidence that ICT investments made by the public sector have resulted in improved performance of teachers and/or learners, nor that it has improved the quality and access to educational resources on the scale predicted.

Nonetheless, there is a general consensus that the ICT opportunity is still valid, and an acceptance as fact that ICT is part of daily life, forever changing the way people learn, work and play. (OECD, 2001, p. 24)

In the British debate following the collapse of the UKeU in 2004, the national funding council explained the lack of interest in e-learning as caused by universities' focus on "blended" learning. A change of focus that was spotted and supported already in February 2003 by Commissionaire Reding in a speech at the opening of the LEARNTEC Forum in Karlsruhe:

Modern e-learning solutions now recognise the importance of learning as a social process and offer possibilities for collaboration with other learners, for interaction with the learning content and for guidance from teachers, trainers and tutors. The learner-centric approaches have put the learner back in command, with a wealth of learning resources at their fingertips. Teachers and trainers once more play a central role, using virtual and traditional face-to-face interaction with their students in a "blended" approach. An approach in which they are no longer seen simply as consumers of pre-determined e-learning content, but as editors, authors and contributors to a contextualised learning scenario. (Reding, 2003, p. 2)

It is important to note that this statement by Reding is not to be read as a total decline of e-leaning or online education, but as recognition of the need for teacher-student interaction, and shared responsibility in the learning process. Neither should the statement be taken as an indication of lost confidence in ICT as the vehicle to realise the vision of the knowledge society by the Commission. The recent Wim Kok report, Facing the challenge. The Lisbon strategy for growth and employment clearly states this:

In order to ensure future economic growth, the EU needs a comprehensive and holistic strategy to spur on the growth of the ICT sector and the diffusion of ICTs in all parts of the economy. The top priority is to implement the eEurope action plan, which calls for measures to promote e-commerce, e-government and e-learning. (European Commission, 2004c, p. 22)

A similar confidence in technology was also stressed in the quotation from the OECD report, but at the same time, the analysis of the situation points toward technology fixation and lack of cultural specificities as the main causes for the lack of success of e-learning:

Technology alone does not deliver educational success. It only becomes valuable in education if learners and teachers can do something useful with it. (...) It is a classic chicken and egg problem. The infrastructure, personal computer penetration and Internet connectivity must be in place for the software and applications to work. On the other hand just having the pipes in place with little or no educational software available generates disappointment and puts many teachers and learners off the online experiences altogether. (...) Another aspect of the 'global chicken and local egg' problem is that educational content does often not travel well across borders. What is appropriate and works in one culture does not necessarily translate to another. (...) educational content and e-learning services (...) need to be tailored to local needs and cultures. (OECD, 2001, pp. 24-25)

Lately, a similar conclusion has been reached in a report on vocational education and training:

e-Learning clearly has the potential to stimulate learning networks and new forms of training organisation. The basic principle of good pedagogy remains that the design of the whole learning process (possibly supported by e-learning) is the decisive factor for the learner's success. Therefore, European countries' e-learning related measures should not be limited to questions of hard- and software, but rather focus on the pedagogy and e-learning in work processes. (European Commission, 2004a, p. 16)

Taken together, the OECD report, the Commissionaire Reding statement, and the VET-report indicate that success of e-learning depends on pedagogical development and a closer integration of technology into the previous learning experiences of students. Technologies to enhance learning should augment the realities of the learning process-interaction, collaboration, and construction–and content should relate to the cultural background of students. A recent report, e-Europe 2005 mid-term review (European Commission, 2004b, p. 5), indicates that "There is a need to systematically evaluate the lessons that have been learnt from all the initiatives and pilot actions in order to set a course for e-learning in the future" (see also European Commission, 2005).

In this chapter, we want to contribute to this process through a refocusing of the learning theoretical framework, and a proposal for future development of e-learning, or technology enhanced learning.

A Learning Theoretical Framework for E-Learning

When the Open University of the UK (UKOU) was set up in 1969, "Not to teach, but to facilitate learning" was chosen as bon mot for the operation. Within the history of education, this signalises a shift of focus from a teacher-centred to a student-centred approach. The role of the teacher changed from content provider (lecturer) to guide and coach of the student's learning process. Also, the concept of learning shifted from knowledge acquisition to knowledge construction.

Without these conceptual changes, the enhancement of learning through the use of technology would never have liberated learning from the constraints of time and place as radically as we experience in these years. The UKOU has continuously explored this track, as educational technology has developed through integration of electronic communication, multimedia, and databases, into an e-learning concept in which tutoring of students plays an essential role, either through face-to-face sessions or via computer conferencing.

The success of the UKOU has often been seen as a direct consequence of their well-prepared educational resources, and the conscious use of educational technology, in a speed adapted to student needs. The point we want to make is that this interpretation overlooks an essential feature in the UKOU concept: the integration of learning activities into the resources and the tutorials, and the ability to make these activities of cultural relevance. In other words, the success of the UKOU is related to the implementation of a social constructivist approach to learning, even though this approach is seldom marketed as such. The difficulties faced by some e-learning and online learning initiatives, mentioned previously, have been caused by viewing learning, and especially e-learning, as a process of knowledge transfer instead of knowledge construction.

Figure 1. Learning objective, learning activities, and learning resources

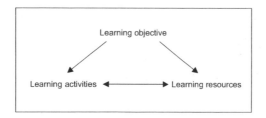

Learning Objectives, Learning Resources, and Learning Activities

In the following, we propose to make a clear distinction between learning objectives, learning resources, and learning activities based on a social constructivist approach to learning (a further elaboration of the social constructivist approach follows later in this chapter). Together, the three concepts describe a learning environment. In a specific learning process, the three concepts cannot be divided, since it is not possible to create a learning environment without any one of them. However, the relationship between the concepts may differ. In Figure 1, the learning objective is placed above learning resources and activities, because they serve the learning objective.

The learning objective describes the purpose of the learning process. What is the student supposed to achieve? What are the educational goals? Which qualifications and competences will the student gain?

Learning resources are materials used to support learning, and consist of content and tools, or other means necessary to reach the learning objective. In recent discussion, learning resources are often called learning objects (a specific discussion of the concept of learning objects follows later in this chapter). Resources are physical things such as books, paper, computers, and so forth, and they only become learning resources when they are viewed in a framework of a learning objective. Learning resources are the "what" in the learning environment: What should the students work with?

Finally, learning activities describe what activities the students should perform in the learning process in order to achieve a learning objective. Learning activities are the "how" in the learning environment: How should the students work with the learning resources in order to reach the learning objective?

The relationship between the three concepts is of importance to the nature of the learning environment, and to the character and role of technology in the learning environment. In principle, two questions may be asked

Figure 2. Learning objective defined within learning resources

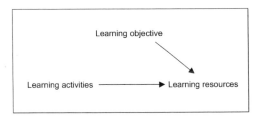

Figure 3. Learning objective defined within learning activities

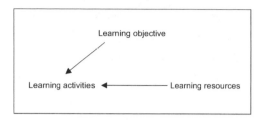

- Should learning objectives be defined within learning resources or within learning activities?

- Should learning resources define activities, or should activities define the use of resources?

Different answers to these questions have different consequences for the nature of the learning environment, and to the understanding of technology enhanced learning.

One possibility is to define the learning objective within the learning resource (see Figure 2). This means that a one-to-one relationship is created — a learning resource covers one, and only one, learning objective. Thus, it becomes possible to make a clear definition of the use of the learning resource, which implies that learning activities are also defined within the learning resource.

Another possibility is to define the learning objective in the learning activities (see Figure 3). This implies that learning activities also define the use of learning resources. Since the use of a learning resource is defined by the learning activities, a learning resource does not necessarily match one, and only one, learning objective. The consequence is that it is not necessary to define within a given learning resource how, and in what context, it is supposed to be used.

The first description of the relationship between learning objectives, resources, and activities is the most widely accepted concept. Nevertheless, in this chapter, we will argue in support of the second one, or for a more balanced view on the relationship between learning activities and learning resources.

An advantage of liberating learning resources from the one-to-one relation with a learning activity and a learning objective is the option of adapting a learning resource to different learning situations by applying different learning activities. Within this optic, the creation of relevant learning activities becomes essential. A learning activity bridges between the interests and cultural heritage of students, and the relevant learning resources chosen to reach the learning objective. Furthermore, the teacher is reinstalled in a position as responsible for organising the learning process by creating the relevant learning activities for his or her students. Later in this chapter, we will argue that this approach supports a social constructivist understanding of learning as an active collaborative process.

Concept of Learning Objects: Pro et Con

Within current discussions of "learning objects," defined as learning resources supplied with a series of agreed metadata (e.g., SCORM), the previously mentioned problematic incorporation of learning objectives is repeated. As stated by Rob Koper (2000), there seems to be an underlying assumption within the field of instructional design and learning objects design that a learning object contains a learning objective, and is designed for one specific purpose:

In distance education and in the classic instructional-design approach it happens fairly often that instructional materials and the media, rather than the learning activities, are central. (Koper, 2000, p. 12)

This is demonstrated by the metaphors used for describing learning objects. Often LEGO™ blocks (toy building blocks produced by the LEGO Company) are used as a metaphor for learning objects. The metaphor is extreme, in the sense that it suggests that learning objects can be assembled in random orders. Instead, David Wiley (2002) suggests atoms as a metaphor for learning objects. Atoms differ from LEGO blocks because they can only be assembled in structures defined by their own structure. The problem, however, with both metaphors is the focus on the resources. Both metaphors assume that a learning environment consists of learning resources, and can, therefore, be assembled by a number or sequence of learning resources.

Nevertheless, the concept of learning objects presents a number of appealing qualities. Learning objects are relatively small objects that are independent of learning context, and that can, therefore, be reused in different courses.

This is the fundamental idea behind learning objects: instructional designers can build small (relative to the size of an entire course) instructional components that can be reused a number of times in different learning contexts. Additionally, learning objects are generally understood to be digital entities deliverable over the Internet, meaning that any number of people can access and use them simultaneously (as opposed to traditional instructional media, such as an overhead or video tape, which can only exist in one place at a time). Moreover, those who incorporate learning objects can collaborate on and benefit immediately from new versions. These are significant differences between learning objects and other instructional media that have existed previously. (Wiley, 2002, p. 3)

The keyword is flexibility. Learning objects are flexible in the sense that they can be reused, rearranged, and resized. To ensure that learning objects can be found by different users, learning objects are ascribed metadata that describe the nature of the object.

As Orill (2002) writes, discussions of learning objects have long focused on the technological possibilities: especially the advantages of reusability.

While there are undoubtedly advantages to the development of these learning objects, we have, as a field, overlooked the most important aspect of the tools — how they support student learning. The discussion on learning objects thus far has focused largely on their design and technical development. (Orill, 2002, p. 2)

Lately, the concept has been the target of massive criticism, especially concerning its proclaimed pedagogical neutrality. As Friesen (2003) argues: "specifications and applications that are truly pedagogically neutral cannot also be pedagogically relevant." However, the next step is not to reject the notion of learning objects, but to begin designing learning objects from a pedagogical approach. Later in this chapter, we discuss how the ideas behind learning objects may be reconsidered in a framework of an active collaborative approach to learning.

Learning as Cooperation or Collaboration?

Changing the focus from learning resources to learning activities also actualises the need for further analysis of the ways students work together in the learning process: Do they cooperate or do they collaborate? The distinction between cooperative and collaborative is made by Roschelle and Teasley:

We make a distinction between 'collaborative' versus 'cooperative' problem solving. Cooperative work is accomplished by the division of labour among participants, as an activity where each person is responsible for a portion of the problem solving. We focus on collaboration as the mutual engagement of participants in a coordinated effort to solve the problem together. (Roschelle, & Teasley, 1995, p. 70)

Cooperation is defined by a well-delineated division of labour. A task is divided into smaller, mutually independent subtasks, and subtasks are distributed among participants who work on their individual task in separation from other participants. This means that participants work on different, and independent tasks with different, or not coordinated, objectives. In principle, the participants do not need to know what the other participants are working with. When all participants have finished their subtasks, the overall task is done as the sum of subtasks.

Collaboration, on the other hand, is a process where the participants take part in a shared task. The participants all work on the same task, and they all have the same objective. Since the participants work on a shared task, they are mutually dependent. In principle, everybody is involved in all the processes and knows what the others are doing. In other words, the participants have a shared and common understanding of the task, as well as the work.

Collaboration is a coordinated, synchronous activity that is the result of a continued attempt to construct and maintain a shared conception of a problem. (Roschelle & Teasley, 1995, p. 70)

Within the field of CSCW (computer supported cooperative work), the emphasis is on *cooperation*, whereas in the related field of CSCL (computer supported collaborative learning), *collaboration* is emphasised. This signalised difference between cooperative work and collaborative learning, more than anything else, probably reflects an intensified development from the industrialised society to the knowledge society over the last 15 years. Productivity has become increasingly dependent on collaborative processes involving learning in problem solving. This is also reflected in the learning theoretical shift from cognitivism to social constructivism in understanding the essence of the learning process[1].

Cognitivism

According to a cognitivist understanding, learning is a process of receiving and processing information. Information is objective and independent of the learner, which means that the cognitivist approach builds on an objectivist concept of knowledge. In other words, knowledge can be objectified. At the same time, the cognitivist approach presents an individualistic understanding of learning. Learning takes place within the cognitive apparatus of the individual by processing of information into knowledge.

Since learning is individual, cooperation is not essential to a cognitive understanding of learning processes. However, the cognitive approach supports cooperation in the meaning "distribution of knowledge." Knowledge can be transferred between individuals. In relation to learning, cooperation becomes a question of knowledge management and knowledge transfer. It is based on the idea of distributing knowledge among individuals working independently.

Social Constructivism

Social constructivism builds on concepts from activity theory (Vygotsky, 1978), claiming that thinking and learning originate in a social practice. As opposed to a cognitivist approach, activity theory regards learning as a social activity situated in a practical context. Further, knowledge is not processed information, but a social construction.

Central to activity theory is the concept of activity, describing the collective activity of humans. Knowledge is tied up within an activity, and learning is an active, object-oriented process that takes place within the social and practical context of human collective activity. Leont'ev (1978) makes a distinction between activity, action, and operation. An activity is guided by an overall motive, actions are directed at goals, and operations point at physical conditions in the world.

Engeström (1987) combines Leont'ev's three dimensions of activity with Bateson's five levels of learning. Engeström defines learning as a development of activity, and further defines Learning I, II, and III as development of

Figure 4. Cooperation and collaboration as different levels of consciousness

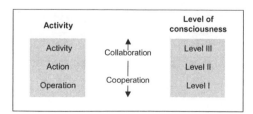

operations, actions, and activity, respectively (Bateson also defines Learning 0 and Learning IV, but they are not included in Engeström's use of Bateson).

Since the motive carries the overall meaning of the entire activity, actions and operations are *indirectly* motivated by the activity. It is not possible for an individual alone to achieve Learning III, but an individual can be *conscious* of the overall activity. Consequently, the three dimensions of activity can also be viewed as three levels of consciousness. It is of importance to a learning process that the individual is conscious of the overall motive of his or her work. The result is a higher level of learning and understanding.

Since an activity is divided into a number of actions, or subtasks, performed by individuals, it is possible to perform an activity through cooperation. Individuals need not be conscious of the overall activity in order to perform it. Thus, cooperative work only requires consciousness of Level I and II. An understanding of the entire activity, however, requires a consciousness of Level III and, therefore, also knowledge of the other individuals' actions, and the motive of the overall activity. As opposed to cooperation, collaboration requires an understanding of, and insight into, the entire activity, supporting a consciousness on Level III.

It is now possible to place cooperation and collaboration in relation to levels of consciousness.

Learning with the consciousness of an overall activity can be achieved through collaboration. But what kind of collaboration does activity theory call for? Activity theory implies a certain approach to collaboration. First of all, learning is not individual, because knowledge is tied up within an activity. It means that learning is not only a question of a collaborative process in itself, but also a matter of participation in a social and practical context. From this point of view, collaboration is not a means to improve learning, but an integral part of the learning process. Central to collaboration, according to activity theory, is (1) participation in a shared activity that is (2) directed at a shared motive.

This leads to the characterization of learning as an *active collaborative process*. Central to learning is not a store of knowledge to be acquired by the student, but rather the activities of the student. This understanding of the learning process is an argument for the approach where a learning objective and the use of learning resources, are determined by activities of the students (cf. Figure 3).

Shifting the Focus to Learning Activities

The question is how active collaborative processes can be supported by technology, and more specifically, how they can be initiated by learning activities. As Wiley (2003) argues, it seems that developers of learning objects, inadvertently, have designed from a cognitivist approach and have, consequently, carried out a one-sided focus on resources:

The instructional methods explicated by many employers of learning objects are surprisingly similar, drawing largely on the same theoretical work from the 1980s or earlier, including work done by Mager (1975), Bloom (1956), Merrill (1983), Clark (1989), and others working in behaviourist or cognitivist instructional paradigms. (Wiley, 2003, p. 1)

Among designers of learning objects there has been a trend towards a "the smaller the better" approach. Earlier in this chapter characterised as the one-to-one approach. The more precisely a learning object is described with a frame of metadata, the better the chance for chasing it and reusing it in another context. Koper (2001) introduces the term "unit of study" as

(...) the smallest unit providing learning events for learners, satisfying one or more interrelated learning objectives. This means that a unit of study can not be broken down to its component parts without loosing its semantic and pragmatic meaning and its effectiveness towards the attainment of learning objectives. (Koper, 2001, p. 3)

Our point of view is that this argument should be brought even further. The loss of "semantic and pragmatic meaning," as identified by Koper, is identical with the "need to be tailored to local needs and cultures," as highlighted in the OECD report as the reason for e-learning's lack of success. We do not oppose the aspect of reusability attached to the concept of learning objects, and we fully support the attempts made to identify clusters of relevant metadata to support search ability in order to place pedagogy in the centre of attention, as proposed by Koper and his colleagues at the OTEC (Educational Technology Expertise Centre at the Open University of the Netherlands) with their educational modelling language (EML).

EML provides the framework to notate and communicate the designs in a complete form, validates them on completeness in structure, makes it possible to identify the functionality of learning objects within the context of a unit of study and provides means for real interoperability and re-usability. (Koper, 2001, p. 27)

The arguments presented earlier in this chapter pave the way for a shift of focus from learning resources to learning activities, in order to integrate the learning process more directly with "local needs and culture" through problem solving activities. Support for this position, we also find in the article by Koper (2001), although we push his views toward the edge by reading them in support of a local and cultural heritage dimension:

(...) a lot of learning does not come from knowledge resources at all, but stems from the activities of learners solving problems, interacting with real devices, interacting in their social and work situation. A lot of research about learning processes provides evidence for this stance that learning doesn't come from the provision of knowledge solely, but that it is the activities of the learners into the learning environment which are accountable for the learning. (Koper, 2001, p. 3)

The challenge is to create technologies that are able to enhance learning in two dimensions: (1) as collaborative learning environments in which learners are able to collaborate and communicate among themselves, and with teachers or tutors, and (2) as interactive learning resources that support learning activities performed by learners alone, or guided by teachers or tutors.

Creating Learning Resources for Multiple Learning Activities

A learning objective directs the work of the students through curriculum requirements, including specification of competences to be acquired. Learning resources form the substance of information from which learning generates through learning activities. These activities may be structured to different degrees, depending on educational maturity of the learner and resources available. At one end of the spectrum, we find structured classroom teaching, which becomes successful, basically, when the teacher initiates collaborative learning experiences. In the middle, research shows that distance education reaches a higher completion rate when collaboration among students is introduced via computer-mediated communication. Similarly, interactive learning resources, in general, enhance the learning outcome through the activities learners are able to perform. At the other end of the spectrum, project-oriented problem-based learning, in which learners collect learning resources collaboratively with guidance from a teacher or tutor (coach), becomes the ultimate self-governed learning process in a collaborative environment close to a research process.

A challenging option in these years is the combination of technology development supplying broader access to high-speed networks, extended use of streaming techniques and increase in storing capacity, and the growing cultural understanding of the need for digitalisation of the cultural heritage. These new databases create opportunities for multiple learning activities to be organised by the teacher according to the needs and learning capacities of his/her students.

Let us try to illustrate what we have in mind by an example. Within a programme for access to cultural heritage, the University Library in Aarhus is digitalising commercials from the twentieth century previously shown in cinemas and on television. The project is supported by the Danish Research Council, and involves researchers from University of Aarhus, utilising this new digital resource in their research. At the same time, access to the material has been negotiated with the Danish copyright organisation (COPYDAN), in order to make the material available for educational institutions in Denmark.

By the end of the project, the database will incorporate approximately 5,000 commercials, dating from 1896 to today. Unfortunately, the first part of the century is not very well represented, since no one really cared for preserving this low-brow culture. But being authentic audiovisual productions, each of the commercials documents and reflects the time in which they were produced.

In an educational context, this resource could obviously be used for teaching advertising strategies over a period of 100 years. But the material also contains sources for tracing cultural developments, for example, how the family — and especially the role of women — has changed, through analysis of commercials for food, cooking, and laundering. Furthermore, an investigation of the development within audiovisual language is an option. The possibilities are numerous. At the same time, the resource offers opportunities for students to learn how to work with large amounts of texts: a competence that is becoming more and more central in the knowledge society. So far, analysis of audiovisual material has been performed on a few examples, from which extensive conclusions have been drawn. With access to large digitalised resources, real quantitative analysis based on qualitative criteria may be conducted.

In order to facilitate analysis of the digitalised commercials in a collaborative learning environment, a tool for annotating comments to each commercial is being developed. These annotations may be stored in a database parallel to the one containing the commercials for continuous expansion, but may also be circulated among collaborating students and researchers.

In the context of this chapter, we want to underline the open nature of such a digital learning resource as the Danish commercials. It is multimedial, containing audio, video, and text, with possibilities of attaching the needed metadata, and it may be activated for a variety of uses. Learning activities to be applied, when the database is used in education, are the ones that bridge between learning objectives (defined by curriculum) and this learning resource. The same resource may be used in different learning scenarios, and for different groups of learners. The task of the experienced teacher, with his or her knowledge of students and curriculum, is to set up or outline a frame for relevant learning activities.

In order to support the creation of learning activities in relation to large learning resources, teaching experiences from former use of the material should be collected in the database, and supplied with a set of relevant pedagogical metadata. Metadata should be dynamic, in the sense that it is possible to constantly attach additional metadata related to the use of the resource, the activities. This way, it should be possible to collaborate between teachers, and to explore the pedagogic opportunities within the learning resource.

As mentioned in the OECD report "Technology alone does not deliver educational success. It only becomes valuable in education if learners and teachers can do something useful with it" (OECD, 2001, p. 24). The challenge is to use technology to enhance learning by increasing flexibility for students and teachers, by offering them tools for collaboration, and by creating options for interaction with large scale multimedial learning resources, through a series of possible learning activities that relate to the social and cultural environment, and bridge to the need of society formulated as learning objectives in the curriculum. Shifting the focus to learning activities also reinstalls the teacher as an educator with responsibility for organising the learning process as a facilitator.

References

Bateson, G. (1972). *Steps to an ecology of mind.* New York: Ballantine Books.

Engeström, Y. (1987). *Learning by expanding.* Helsinki: Orienta-Konsultit Oy.

European Commission (2000). *E-Europe—An information society for all.* Retrieved March 1, 2005, from http://europa.eu.int/scadplus/leg/en/lvb/l24221.htm

European Commission (2003). *E-learning: Designing tomorrow's education.* Retrieved March 1, 2005, from http://www.elearningeuropa.info/extras/pdf/mid_term_en.pdf

European Commission (2004a). *Achieving the Lisbon goal: The contribution of VET.* Executive Summary. Retrieved March 1, 2005, from http://europa.eu.int/comm/education/policies/2010/studies/maastrichtexe_en.pdf

European Commission (2004b). *E-Europe 2005 midterm review*. Retrieved March 1, 2005, from http://europa.eu.int/geninfo/query/engine/search/query.pl

European Commission (2004c). *Facing the challenge. The Lisbon strategy for growth and employment*. Retrieved March 1, 2005, from http://www.eu-oplysningen.dk/upload/application/pdf/f9a0512d/wimkok.pdfOECD(2001): *E-learning: The Partnership Challenge*. Retrieved March 1, 2005, from http://www.oecd.org/LongAbstract/0,2546,en_2649_33723_1898362_1_1_1_1,00.html

European Commission (2005). *Challenges for the European information society beyond 2005*. Retrieved March 1, 2005, from http://europa.eu.int/information_society/eeurope/2005/index en.htm

Friesen, N. (2003). *Three objections to learning objects and e-learning standards*. Retrieved October. 25, 2004, from http://www.learningspaces.org/n/papers/objections.html

The Guardian. (2004, April 13). *The online revolution, mark II*. Retrieved March 1, 2005, from http://education.guardian.co.uk/egweekly/story/0,,1190372,00.html

Koper, R. (2000). *From change to renewal: Educational technology foundations of electronic environments*. Retrieved February 23, 2004, from http://eml.ou.nl/introduction/articles.htm

Koper, R. (2001). *Modeling units of study from a pedagogical perspective: The pedagogical meta-model behind EML*. Retrieved February 23, 2004, from http://eml.ou.nl/introduction/articles.htm

Leont'ev, A. N. (1978). *Activity, consciousness, and personality*. Retrieved February 2, 2005, from http://www.marxists.org/archive/leontev/works/1978/index.htm

OECD (2001): *E-learning: The Partnership Challenge*. Retrieved March 1, 2005, from http://www.oecd.org/LongAbstract/0,2546,en_2649_33723_1898362_1_1_1_1,00.html

Orill, C. H. (2002). Learning objects to support inquiry-based, online learning. In D. A. Wiley (Ed.), *The instructional use of learning objects*. Retrieved June 15, 2004, from http://www.reusability.org/read/

Reding, V. (2003). *Is e-learning going mainstream*? Opening of the Learntec Forum, Karlsruhe, February 4, 2003. Retrieved Sept. 14, 2003, from http://europa.eu.int/rapid/start/cgi/guesten.ksh?p_action.gettxt=gt&doc=SPEECH/03/48|0|AGED&lg=EN&display

Roschelle, J., & Teasley, S. D. (1995). The construction of shared knowledge in collaborative problem solving. In C. O'Malley (Ed.), *Computer supported collaborative learning* (pp. 69-97). Heidelberg: Springer-Verlag.

Vygotsky, L. S. (1978). *Mind in society.* Cambridge, MA: Harvard University Press.

Wiley, D. (2002). Connecting learning objects to instructional design theory: A definition, a metaphor, and a taxonomy. In D. A. Wiley (Ed.), *The instructional use of learning objects.* Retrieved June 15, 2004, from http://www.reusability.org/read/

Wiley, D. (2003). *Learning objects: Difficulties and opportunities.* Retrieved August 17, 2004, from http://wiley.ed.usu.edu/docs/lo_do.pdf

Endnote

[1]   Both "cognitivism" and "social constructivism" are terms that cover a range of different theories. In this chapter, however, we use the terms only to describe generalised theoretical positions or paradigms.

Chapter IX

Computer Supported Collaborative Learning and the Central Research Questions:
Two Illustrative Vignettes

Tony Carr, University of Cape Town, South Africa

Vic Lally, The Mental Health Foundation, UK & University of Sheffield, UK

Maarten de Laat, University of Southampton, UK

Glenda Cox, University of Cape Town, South Africa

Abstract

This chapter examines the theoretical and conceptual issues involved in gathering evidence to build a database for the design of virtual higher education (computer supported collaborative learning — CSCL — and networked learning — NL). After briefly surveying the current state of CSCL/NL research and its lack of theoretical synthesis, we propose three high-level research questions as a way of focusing our efforts on finding answers. We then offer two vignettes of empirical case studies from our own research. These studies are used to illustrate the challenges to be faced, and possible approaches to be used, in addressing the questions. In particular, we look at the way theory and praxis (theory-informed practice) might be more effectively engaged through "theory-praxis conversations," in order to make effective use of empirical data in building the evidence base that will be needed to design and build virtual higher education over the next 10 years.

Preamble

In the brief for this book, we were asked to be bold in our ideas, challenging in our approach, grounded in the literature, and to be creative in ways that will enhance learning. This is a tall order indeed. We are pleased to be asked to address this challenge, but we will leave it to others to decide how we have measured up to its demands. Because we are engaged in education, we are accustomed to this level of challenge: attempting to enhance learning through technology presents researchers and teachers with many such challenges. The context for the ideas and research presented here is a form of education referred to as computer supported collaborative learning[1] in the U.S. and networked learning in the UK and Europe (to some extent). In particular, we will focus on CSCL/NL research and praxis (theory-informed practice) using small examples (vignettes) of data from higher education settings in the UK and South Africa. Before proceeding, it may be useful to say something about two underlying ideas in our work. The idea of "evidence" is central to our educational thinking in this work. To be able to draw explicitly on an international and publicly available body of understanding that informs the design of education is, we think, one of the important ways in which we may hope to systematically improve the quality of CSCL/NL education through research. One of the high-level challenges we face is to build such an evidence base for CSCL/NL designers and educators to draw upon. Readers who are interested in pursuing this idea of evidence and education, in more depth, might also refer to the excellent account provided by Elliott (2001). Some of the central research questions surrounding the building of the CSCL/NL education evidence base form the main focus of this chapter. Using the evidence base efficiently in the educational design process is a related challenge (not discussed here), and details of some of these issues may be found in the work of Goodyear and others (Goodyear, 2001; Goodyear, Avgeriou, Baggetun, Bartoluzzi, Retalis, & Ronteltap, 2004a; Goodyear, De Laat, & Lally, 2005).

The second important underlying idea is that of education. We see education as an act of ethical intervention in our teaching and learning processes, to enable groups or individuals to pursue desirable aims and purposes more effectively, more coherently, more humanely. This presupposes that as active, enquiring agents, humans are continually and authentically engaged in these processes, both consciously and unconsciously, in ways that have an ethical basis (an ethics about which we can conduct a reasoned argument — see Taylor (1991) for more details). These teaching and learning processes are in a dynamic state of development, and also a central feature of all our actions, whether intentional or otherwise (we may not "control" them entirely). One of the central aims for the CSCL/NL educator then, arising from these ideas, is to intervene, in educational contexts, in the development of these processes, both at the individual level and among groups, in order to assist participants in a more effective pursuit of their own desirable and ethical purposes.

Intervening, in a planned way, in the development of CSCL/NL teaching and learning processes, to assist participants in the pursuit of their own wider aims and purposes, is the central purpose of educational design for CSCL/NL. A design is a plan, template, or framework for educational intervention. This may seem relatively straightforward. However, it is complicated by at least six factors. Teaching and learning processes (between and within individuals) are dynamic: they change over time, before, during and after educational events. They interact with each other: individual and group processes interact, teaching and learning processes interact; all of this may change the cogency of any intervention. They are creative, and can develop in unexpected and unanticipated ways that are also desirable, even though they could not easily be planned. They are mediated by affective and social processes, and by language and other tools that have both affordances and limitations. They are modified by refection before, during, and after an educational event. They are subject to reflexive action to some degree: that is, they are subject to the motivation, aims, purposes, and desires of the active human agents who engage in them. This is not an exhaustive list. It serves to indicate that educational design is not a trivial undertaking; it is attempting to nurture a shared direction for the development of teaching and learning processes, while acknowledging that these factors are too complex to be controlled, leaving aside the ethical implications of attempting to exert such control. Nor can they be investigated by methods that rely upon the controlling of variables. Other qualitative methods are required (we will refer to this later).

Arising from this is a contingent complexity in the nature of educational design for CSCL/NL. It is, of course, not possible to address all these factors within one chapter. A major challenge for educational designers is to think about, and enact design, in time-efficient ways that can be of practical help to teachers, in the face of this complexity. One of the challenges is to develop an evidence base derived from theory and praxis that can be used to guide the design of CSCL/NL.

This chapter will now to try to formulate some high-level research questions for those trying to build the evidence base for CSCL/NL design. We will follow this by providing vignettes of results, and analysis from two recent empirical studies, to illustrate possible approaches to addressing these questions.

CSCL/NL, as an area for serious research, is only now on the edge of achieving programmatic coherence. A little history might help us to make this point, offer the CSCL/NL research community a perspective on our current situation, and help to refine our focus for the future. The 1997 CSCL conference in Toronto certainly had a great measure of ideas in the mix (Ontario Institute for Studies in Education, 1997). But it was a "mix". The conference invitation referred to design issues, methods of instruction, theoretical issues of learning and collaboration, specific pedagogical models, assessment, and "research" in technologically-mediated communication. By 2002 (Stahl, 2002), the centrality of human learning had emerged as a pre-eminent focus at the CSCL conference in Boulder CO. Another eighteen months on, and the CSCL 2003 conference, in Bergen, consolidated this focus (Wasson, Ludvigsen, & Hoppe, 2003), but also moved us on with a strong emphasis on pedagogical design of learning environments, and on theory and theoretical frameworks for design. There were also equally important parallel emphases on improvements and innovations, and the analysis of collaborative activities. The lesson that we are drawing from this brief history is that we have quickly moved from a broad and rich mixture of interests, with little coherent focus, to a more sustained focus on learning, with interests in theory and practice running in parallel tracks, although these are not strongly related.

CSCL/NL is an increasingly important area for rigorous academic study, as significant effort and resources are deployed to develop and implement sophisticated pedagogic designs in schools and universities. But, as yet, it is without a theoretically informed, coherent research evidence base. The CSCL/NL research community is relatively well placed to undertake the task of building this evidence base and bringing coherence to it. However, CSCL/NL is a relatively new field of research endeavour (Goodyear, Banks, Hodgson, & McConnell, 2004b, p. 3) in which there is a relatively small body of empirical research. Much of this work is based upon one or the other of a range of theoretical perspectives that have been employed (Paavola, Lipponen, & Hakkarainen, 2002). The field of CSCL/NL research, we conclude, is theoretically fragmented. At the same time, as we have already pointed out, theory and practice are not strongly related.

We have argued elsewhere (De Laat & Lally, 2003) that CSCL/NL is a new "interdiscipline" that is drawing on theoretical perspectives from cognitive and social psychology, education, computing, and applied linguistics. We use the term "interdiscipline" to describe this multitheoretical situation. Johnston has argued that the rate of change in the evolution of universities is exponential, and that we are in now in the age of integration and synthesis (Johnston, 1998). Disciplines, she argues, are under enormous pressure to cross boundaries, to synthesize and integrate in order to move forward their research agendas. For CSCL/NL research, and other new interdisciplines, we argue that this is both a creative and a problematic process. It is creative because researchers and practitioners are, as we have witnessed at CSCL/NL conferences, bringing a range of exciting theoretical perspectives to bear upon and illuminate the field, and engaging in a wide range of new pedagogical approaches and designs. It is problematic because these perspectives and designs are, as yet, partial, fragmented, and weakened by lack of synthesis. We propose that synthesis is currently one of the key challenges for CSCL/NL as an interdiscipline. This chapter, therefore, is also partly concerned with what we term "theory-praxis conversations," and how they may be used to address this challenge, and enhance the educational evidence base for CSCL/NL. We are attempting a further development of our argument: that theories informing CSCL/NL and the praxis of CSCL/NL (by praxis we mean practical activity in real CSCL/NL courses that is in some way informed by theoretical perspectives) can "converse" in important ways, and this may be used to enhance the quality of our "evidence base" for CSCL/NL design. These conversations may take place in research and practitioner communities.

We contend that through these conversations, we may address three, central, high-level research questions that we suggest are facing CSCL/NL over the next 10 years:

- How do we achieve a synthesis of multitheoretical perspectives, and thereby, increase their power as a tool in the research process (Halverson, 2002), and as a guide to the design of praxis?

- How do we take account of the richness and complexity of praxis, and understand the extent to which it is currently "beyond theory" in any case or context?

- How do we support the creative interaction of theory and praxis, and the research methodologies used to facilitate this?

This chapter provides vignettes of two empirical case studies of CSCL/NL that focus on discussion within CSCL/NL activities. Discussion between participants is a central feature of CSCL/NL. During CSCL/NL activities participants are stimulated to interact and collaborate with each other to fulfil and coordinate their learning needs. In research terms, much is still unclear about the most effective forms of CSCL/NL. As we have argued, there is a need for research and development of new understandings that will provide guidance on the design and moderation of CSCL/NL. Stahl (2003) takes this point further by explicitly arguing for a more appropriate conceptual framework and analytic perspective to guide this work. At present, he suggests we are witnessing an emerging conceptualisation, where concepts borrowed from other theories and philosophies are being adapted, but as yet, we still lack a sufficiently powerful theoretical base to guide our research and our praxis. This is increasingly acknowledged as a concern among researchers in the field, and was clearly expressed during the CSCL 2003 conference in Bergen (Beuschel, 2003; Hakkinen, Jarvela, & Makitalo, 2003; Stahl, 2003; Wasson et al., 2003). The need for more empirical research to provide an evidence base for this emerging conceptual framework is clear. We think it is important that this research is focused on the central processes of CSCL/NL, that is, learning and tutoring. We believe that these understandings will contribute to the development of better pedagogical frameworks and software that more effectively support learning and tutoring by design.

This study is a continuation of our investigations into learning and tutoring processes occurring in online (virtual) communities of students engaged in CSCL/NL. Previously, we focused, informed by constructivist and sociocultural perspectives, on content (discourse) analysis (CA) of learning and tutoring behaviour, sometimes in combination with the use of critical event recall (CER), to probe the tutor's account of his management and facilitation of the processes involved. In this paper, we focus on the students' and tutors' behaviour, using the same research methods, in two contexts (synchronous and asynchronous), in order to provide a more holistic description of two CSCL/NL communities. This work is guided by our previous argument: that both the university tutor and the learners contribute fully to the organisation and regulation of their learning event. As such, every member of this community may be seen as both learner and tutor. Of course, the designated tutor continues to have a *status apart,* being responsible for the overall coordination of the workshop and its educational goals. But during the learning tasks, the tutor acts more as a "guide on the side," moderating, stimulating, and learning by taking part as a coparticipant in the online discussions.

Effective participation in CSCL/NL requires the development of appropriate communication, coordination, and regulation skills. At the same time, we must be cognisant that other aspects of individual human agency, such as motivation, identity, and social presence and awareness, are significant variables in any educational context, and affect the possibility of meaningful and balanced online discussions. Constructive group interaction and dynamics also involve positive interdependence (group belonging and the awareness that each member's effort is important for the group success) and individual accountability (each participant's contribution is valued and balanced in the collaborative learning process). It is clear that in an educational setting, the development of these complimentary and necessary dynamics cannot be left to chance. Awareness of key role behaviours and strategies is important for the tutor to manage and sustain healthy group dynamics. Participating in CSCL/NL is also demanding for the learner, requiring the development of awareness of his/her role in the instructional process and, in more advanced educational contexts, to take over some of the managerial responsibilities for the development of the discussion.

Using content analysis and critical event recall interviews with the learner coparticipants, we attempt to provide relatively rich descriptions of how CSCL/NL processes are coordinated and regulated among them. Hakkinen et al. (2003) suggested a multimethod approach that is process oriented, and takes into account different contextual aspects of CSCL/NL. They argue that research is needed that captures the process and organisation of collaborative interaction and its contribution to learning:

Methods should be developed not only for capturing processes and outcomes of learning, but also experienced effects and individual interpretations of participation in CSCL settings. (Hakkinen et al., 2003, p. 402)

The aim of this kind of research is to provide a more complete picture of CSCL/NL processes, and to contribute to more profound analysis of virtual interaction. We suggest that this is in no way an easy task; CSCL/NL is a complex domain of educational endeavour for researchers and participants.

Samples and Methods

Sample for Case Study 1

The participants featuring in this case study were undertaking a Master's Programme in E-Learning. This M.Ed. programme is based upon the establishment of a "research learning community" among the participants and the university tutor. It is fully virtual; there is no scheduled face-to-face contact in the 2 years of the part-time programme. In this, community activities are undertaken around five "workshops" over a 2-year period. The programme is hosted in the virtual learning environment, WebCT. The students are mainly midcareer professionals, many of whom have postgraduate experience of higher education, are themselves professionally engaged with teaching responsibilities, and are often charged with developing e-learning within their own organisation. Our analysis is based upon collaborative project work conducted by seven students and one tutor in the first workshop of this programme (approximately 10 week's duration).

Sample for Case Study 2

For this case study, we used samples of the first 100 statements from each of three unmoderated online "negotiating" chats in a one semester, blended, trade-bargaining simulation. This course was offered as a third-year undergraduate economics module to students at the University of Cape Town, and was studied by 98 students in 2001. Students take on the role of trade delegates from specific countries, and are assigned to negotiate trade motions in particular areas for example, agriculture, trade, and intellectual property. Participation in the simulated World Trade Organisation bargaining round is driven by intrinsic and assessment incentives for collaborative and competitive behaviour, and contained by a carefully structured process for the agenda setting and bargaining rounds of negotiation. Online communication in the form of chats, online discussions, and the sharing of online resources provide spaces for interaction and coordination. It also facilitates the management of emerging knowledge, and access to a resource base. Our analysis is based upon trade negotiations, conducted by 11 students at three different points in the semester.

Methods for Case Study 1

Content Analysis

The central purpose of content analysis is to generalise and abstract from the complexity of the original messages, in order to look for evidence of learning and tutoring activities. In order to probe CSCL/NL (learning and tutoring), we "coded" the contributions using two coding schemas. The first coding schema, developed by Veldhuis-Diermanse (2002), was used to investigate the learning activities in the group. This schema includes four main categories: cognitive activities used to process the learning content, and to attain learning goals; metacognitive knowledge and metacognitive skills used to regulate the cognitive activities; affective activities, used to cope with feelings occurring during learning; and miscellaneous activities. We decided to exclude the miscellaneous category in our analysis, since we are interested in the evidence of learning activities. To focus on tutoring activities in the group, we used another coding schema (Anderson, Rourke, Garrison, & Archer, 2001). This schema includes three main subcategories: design and organisation, facilitation of discourse, and direct instruction. Our intention here was to attempt to reveal the ways in which the participants were facilitating and regulating each other's learning, while undertaking the workshop project task.

In order to make the CA task manageable, we sampled the message data from the workshop (approximately 1,000 messages were posted during the task). We divided the 10-week period into three sections: beginning, middle, and end. From each period, we took a 10-day message sample to form our data set. In each sample, we analysed messages in selected threads, rather than sampling across threads. This was important to enable us to follow and code the development of learning and tutoring within an ongoing discussion, rather than across unrelated messages. This resulted in a selection of 160 messages. Codes were assigned to parts of messages based on semantic features such as ideas, argument chains, and topics of discussion (Chi, 1997). Capturing these activities using strict syntactic rules was not possible because of the elaborate nature of much of the discussion. We chose to use NVivo software to help us to partially automate this process: to highlight segments of the text with coding that, we claim, represents a particular learning or tutoring activity. In effect, these coded segments were our units of meaning. NVivo was also used to conduct searches of the coded data, in order to produce summary tables (see Table 1). We used the following procedure to determine intercoder reliability. Firstly, for each coded message, we checked to see if the codes assigned by the two coders referred to the same parts of the message (i.e. the same units of meaning). Secondly, we checked to see if the two coders had assigned the same codes to each unit. Based on a 10% sample of all the messages coded by the two researchers, a Cohen's Kappa of 0.86 was established. This indicates an acceptable level of agreement between the two coders.

Critical Event Recall Interviews

Content analysis has provided us with evidence of learning and tutoring process patterns that were occurring in this group during the workshop task. To understand these patterns further, we used the summary results of the CA as a stimulus for critical event recall interviews with the participants. This was done to gain feedback from them about their own understandings of the patterns that emerged, and to help us to understand the context in which these patterns were emerging. The CER interviews enable the articulation of many previously unexpressed aspects of learning, and help to contextualise and elucidate individual behaviour, based on personal motives and perceptions, in relation to the task and the other participants. Therefore, we pursued those situational and contextual aspects of CSCL/NL that were identified by participants during these recall interviews. The interview layout contains two parts. The first part is based on stimulated recall of the learning event. During the second half of the session, the opportunity for post hoc reflections is provided, with additional follow-up questions to help probe and understand the group processes. We have adopted two approaches guiding the CER interviews. Firstly, the participant is presented with a summary table of individual learning and tutoring results for all phases of the discussion (Table 1). Secondly, the full text of the workshop discussions, available in WebCT, was used to recall learning events. The results of the recall then provide the base for the post hoc reflections interview. The selection of the participants for the recall interviews were based on the patterns represented in Table 1. The recall interviews (with an average time of 75 minutes) were transcribed and analysed by the researchers, together.

Methods for Case Study 2

Case study two draws on, and extends, previous research on this course reported by Carr, Cox, Eden and Hanslo (2004) that considers the gendered nature of chat communication within the frame of learning in a community of practice. We also draw on Cox, Carr, and Hall (2004), who consider the effectiveness of chat as a mode of learning communication in two courses. Both of these articles use exchange structure analysis (Kneser, Pilkington, & Treasure-Jones, 2001). In this case, however, we use the schemas of learning and teaching processes employed in case study one to analyze both overall behaviour across three chats, and changes relating to phase shift and group dynamics. We made use of Excel™ spreadsheets to record and analyze the coding. This proved adequate for our needs because the brevity of the chat messages meant that each message could be treated as a single unit of meaning. The research process involved the combination of several research methods, including lab and classroom observations, analysis of chat logs and videotapes of classroom interaction, student and educator interviews, and an evaluation survey of a sample of students.

We found that coding with the schemas for teaching and learning processes proceeded very efficiently, however, we experienced some interesting challenges in the application of the affective category. The schema of learning processes seems to assume that learning is either cognitive, metacognitive, or affective. However, our observation of chats suggests that much of the conversation has an overtly affective component. This is consistent with work by Damasio (1996), who argues that the brain structures responsible for emotion and reasoning overlap, to the point where the ability to experience feelings is often a prerequisite for effective reasoning. The other issue that presented challenges was the intense layering of conversation and metaconversation within the trade bargaining simulation. Conversation included directly negotiating agenda items and trade motions, discussing the best strategies to achieve a successful trade rule, World Trade Organisation procedures, and even the assessment system for the course.

Results and Discussion

In this section, we present the results of analyzing the two case study contexts.

Case Study One: Workshop One of the Master's Programme in E-Learning

Tables 1 and 2 show the results of coding for learning and tutoring processes in the beginning phase of the activity (the first 10 days) for Seline's group (n=10 including tutor) (De Laat & Lally, 2004; Lally & De Laat, 2002, 2003).

In Table 1, we can see that 23 (30%) units were coded for cognitive statements, and Sabine and Monique are mostly involved with this activity. Metacognitive processes are the largest category (31), representing 41% of all coded units. At this stage of the activity, the group is still in the process of establishing relationships as a working group, and trying to understand and conceptualize the collaborative task it is about to undertake. This is a pattern of learning process activity that we have encountered in other groups at this stage in this programme (De Laat & Lally, 2004; Lally & De Laat, 2002, 2003). Tutoring processes, as indicated by coded units of meaning, are dominated by facilitation at this stage of the activity (see Table 2: 63 units [53% of total]). Considerable group processing is also devoted to instructional design (44 units, 37% of total), as members of the group help each other to be organized for the task ahead.

In Tables 3 and 4, we can see results of applying the same coding analysis to identify learning and tutoring processes in Nitin's group (n=8 including tutor) for the same (beginning) phase of activity. An interesting difference occurs when we look at the relative proportions of metacognitive and cognitive processes. They are reversed, compared to Group 1 (9.3% and 17.1% respectively), with more cognitive processing than metacognitive activity going on in Group 2. This

Table 1. Units of meaning coded for learning processes during the beginning phase of Group 1 (case study 1)

Beginning phase	Learning processes of individual community members									
Type of learning Process	Seline*	Sabine	Calvin	Monique	Pierre	Amani	Johann	Kiel	Jaquita	Alan	Total
Cognitive	0	7	2	7	3	3	0	1	0	0	23
Affective	2	6	5	5	2	2	0	0	0	0	22
Metacognitive	0	6	10	1	3	9	0	2	0	0	31
Total	2	19	17	13	8	14	0	3	0	0	76

suggests that the group has engaged with the task itself at an earlier point than Group 1. In terms of tutoring processes in Group 2, a similar pattern is found to that of Group 1 (Table 4).

In the beginning phase, it is evident that except for Mort, all the participants are in some way engaging with learning activities, conceptualising the task ahead. Danton, Mary, and Anka are the strongest contributors. One can say that all activities are equally spread over the participants. Among the members, there is a strong emphasis on making cognitive (46%) statements to each other (debating ideas, using and linking new information). Twenty-nine percent of the coded units are affective, that is, emotional responses to options of other students' task content, and 25% of the codes refer to metacognitive units. It is interesting to see that Brian, the teacher, is present with respect to the learning processes.

An interesting difference occurs when we look at the relative proportions of metacognitive and cognitive processes. They are reversed compared to Group 1 (9.3% and 17.1% respectively), with more cognitive processing than metacognitive activity going on in Group 2. This suggests that the group has engaged with the task itself at an earlier point than Group 1. In terms of tutoring processes in Group 2, a similar pattern is found to that of Group 1 (Table 4).

Table 2. Units of meaning coded for tutoring processes in the beginning phase of Group 1 (case study 1)

Beginning phase	Tutoring processes of individual community members										
Type of tutoring process	Seline*	Sabine	Calvin	Monique	Pierre	Amani	Johann	Kiel	Jaquita	Alan	Total
Direct instruction	7	0	1	0	0	3	0	0	0	0	11
Facilitation	6	7	14	10	3	13	5	5	0	0	63
Instructional design	3	5	10	4	1	17	0	4	0	0	44
Total	16	12	25	14	4	33	5	9	0	0	118

Table 3. Units of meaning coded for learning processes during the beginning phase of Group 2

Beginning phase	Learning processes of individual community members							
Type of learning process	Brian*	Ryan	Mary	Danton	Anka	Aimi	Neem	Mort	Total
Cognitive	5	2	4	4	6	1	2	0	24
Affective	3	1	4	4	3	0	0	0	15
Metacognitive	1	0	4	2	3	2	1	0	13
Total	9	3	12	10	12	3	3	0	52

Table 4. Units of meaning coded for tutoring processes in the beginning phase of Group 2 (case study 1)

Beginning phase	Tutoring processes of individual community members								
Type of tutoring process	Nitin*	Ryan	Mary	Danton	Anka	Aimi	Neem	Mort	Total
Direct instruction	7	0	2	0	5	5	0	0	19
Facilitation	11	11	11	5	15	8	3	0	64
Instructional design	9	8	12	1	15	8	1	0	54
Total	27	19	25	6	35	21	4	0	137

Facilitation is the main activity (64 units of meaning, or 46.7% of the total), although slightly lower as an overall percentage than Group 1. Instructional design processing is at a very similar level to Group 1 (39.4% or 54 units).

Turning now to the CER interviews, the following represent a very brief sample of the rich details available in these. In this sample, we focus on the different levels of experience of the two tutors, and the effect that this had upon their respective groups.

Group 1

This group was tutored by a faculty member who was new to the role. She was going through a period of adjustment to the nature of the task.

Seline (tutor):
In the first few weeks I found the role very intimidating...
Everything happened so quickly...
I had a group of extremely confident people...
I was not sure what my role was...
It helped me to go to the other groups and see what they were doing...
The momentum was driven by the most sophisticated contributors...

This student reflects the effect that this had on the way the group was feeling during this early period:

Sabine (student):

I don't think we realized the enormity of the project.I wasn't feeling very clued-in as to what we should do...

I felt that the atmosphere wasn't particularly good...

Group 2

The tutor in this group had much more experience of CSCL/NL. He knew when to intervene, and when to let the group support itself as it progressed with its task.

Nitin (tutor):

I tried to help them explore the process and identify a project, by giving them an advance organizer (a la Ausubel).

They only really picked up on the content.

They were coming up with lots of ideas.

I noticed the implicit complexity and wanted to sound a note of caution.

I knew they were going to have to come to the point about process.

They needed to decide what the nature of the product will be.

It is evident that this group quickly became task-focused and began to develop its project.

Aimi (student):

Anka was the driving force behind us...

We were trying to organize the roles...

Nitin didn't have a high profiler anywhere, but when he came in he was helpful.

'There was a lot of social chit chat initially, it was a way of...(interviewer: there was some engagement before starting the business?) Yes'

Anka and I phoned each other'

Case Study Two: The Trade Bargaining Simulation

We sampled three chats by a negotiating subcommittee (11 students) that were distributed across the period of the simulation. Chat 1 was their first negotiating chat, while Chat 2 took place soon after the start of the bargaining on trade motions started. Chat 3 was close to the end of the simulation. The first 100 statements of each chat were coded for teaching and learning processes. This sample of 300 statements, drawn from three chats, is just under a quarter of our dataset for this group of 1,334 statements across eight chats (Table 5).

From lab and classroom observations, and student and educator interviews, we are able to make a few observations about this group that may provide some context for analyzing their teaching and learning processes. This group took several weeks to start making sense of the purposes and process of the simulation, and was very prone to highly emotive eruptions of conflict driven by personality clashes. The group was also poorly led, in the sense that the student who dominated the chats, in terms of volume of statements across both learning and teaching schemas, was also the focal point of sporadic conflict (often to the point of flaming). This group's performance was not helped by a tutor who provided limited, and sometimes confusing support, in face-to-face meetings.

The largest share of messages is coded as cognitive statements and we regard this as positive because it indicates a group that is working to achieve shared and individual outcomes. The high percentage of metacognitive statements relates to critical involvement with both the content and process of trade bargaining. The "affective" category undercounts the percentage of messages with a strong community building or affective component because many statements coded as cognitive or metacognitive also had a clear emotive element. This tended to take

Table 5. Units of meaning coded for learning processes for all chats (%) (Case study 2)

Learning processes of chat group members for all chats	Total (%)
Cognitive	43.92%
Affective	25.34%
Metacognitive	28.04%
Rest	2.70%
Total	100.00%

Table 6. Units of meaning coded for learning processes for individual chats (Case study 2)

Learning processes of chat group members for individual chats	Chat 1	Chat 2	Chat 3
Cognitive	33	41	56
Affective	31	31	16
Metacognitive	31	24	28
Rest	5	4	0

forms such as humour, sarcasm, or anger. However, the choice to avoid double coding meant that these aspects were obscured.

Overall, as the simulation proceeds, students become more engaged with the process and content of the negotiations. Affective statements decline, and cognitive statements increase sharply. Metacognitive statements fluctuate slightly across the three samples.

In Table 6, we can see the results of coding individual chats for learning processes. Chat 1 was the opening chat for this subcommittee. Participants needed to resolve their own confusion and uncertainty concerning the process of the agenda-setting phase of the simulation, and about the process of negotiating in a chat room. Much of the conversation was of a very social nature, and by the end of the 100-message sample, analysed here, the participants had still not chosen a chairperson for the discussion. Even deciding what should be discussed presented problems. The strongly affective component of the chat (31 statements) was partly an indication of the level of discomfort experienced by participants in their first chat of the simulation.

Chat 2 took place in the early stage of the bargaining phase of the simulation. One would expect some problems as this was soon after a phase shift, but the coding does suggest that Chat 2 was more effective than Chat 1. The increase in cognitive statements is consistent with a tendency towards a more elaborated argument for and against proposed trade motions. The slight decrease in metacognitive statements is consistent with a group that has achieved more certainty about the process of the chats and of the simulation as a whole. The proportion of affective statements remains high, as personal conflict between group members is about to reach a head. (If we had continued coding beyond the first 100 statements, then an overt outburst of rather nasty flaming would have further increased the role of affective statements.) If we consider the stages of group development, then this chat definitely shows the group reaching "storm-ing" mode.

Chat 3 took place towards the end of the bargaining round. Students knew that the last date for passing motions through plenary meetings was fast approaching, so they were very focused on effective negotiations that could produce a stream of sound proposals with widespread support to the subplenum meetings. The proportion of cognitive statements continues to rise, as students engage more analytically with the discussion of proposed motions. The role of affective statements declines by a similar proportion, and this further underlines the increased engagement of students with on-task communication. The slight increase in metacognitive statements results from several statements about the detailed application of the WTO bargaining procedure.

In total, 182 of the 300 statements (60.6%) were coded as tutoring processes. This is quite encouraging, since it confirms the assumption that students will

Table 7. Units of meaning coded for tutoring processes for all chats (%) (Case study 2)

Tutoring processes of chat group members for all chats	Total (%)
Instructional design and organisation	26.92
Facilitating discourse	20.33
Direct instruction	52.75
Total	100.00

Table 8. Units of meaning coded for tutoring processes for individual chats (Case study 2)

Tutoring processes of chat group members for individual chats	Chat 1	Chat 2	Chat 3
Instructional design and organisation	16	22	13
Facilitating discourse	17	12	12
Direct instruction	20	43	51

exhibit tutoring processes when involved in a process with clear objectives and incentives, but no overt facilitation by educators.

Direct instruction tended to predominate, as students increasingly argued and presented the merit of their proposed trade motions, in detail. This tendency grew stronger towards the end of the simulation. The role of "Instructional design and organisation" was consistent with the many levels of metaconversation inherent to the trade bargaining simulation. The strand of facilitative statements, which ran through the chats, suggested that the students had learnt to take on a facilitative role in the absence of tutor facilitation.

Overall, the increase in statements related to direct instruction over the three chats indicates a growing focus on task-related communication (Table 7). Fluctuations in statements coded for instructional design are related to phase shift and student understanding of the process at the levels of the chat and the whole simulation. The decline in statements that facilitate discourse is also a likely consequence of increased student familiarity with the process and procedure, and improved motivation, by the time of the third chat. Chat 1 is ineffective as a negotiating forum; this is shown in the fact that only just over half of the postings could be coded with the teaching schema. This chat barely got off the ground due to disputes concerning the appropriate topic of conversation. The sample from this chat suggests that the initial confusion and uncertainty about process was compounded by a lack of clear and legitimate leadership. Some of the more productive interactions during this chat are shown in the coding sample in Table 9.

Table 9. Sample of coding for Chat 1

Statement no	Student	Statement	Learning	Tutoring
1	Student one	I have proposal to bring forward on coral reef protection in [my country]	metacognitive	direct instruction
2	Student two	that sounds interesting but can you elaborate	cognitive	facilitating discourse
3	Student three	Proposal on protection on coral reef and doubling up on the marine life?	cognitive	facilitating discourse
4	Student One	Yes	cognitive	
5	Student three	We could look at the marine life aspect too, because there is that relation.	metacognitive	direct instruction
6	Student four	question - what impact does the protection on coral reefs has on the word trade - any barriers	cognitive	direct instruction
7	Student two	to avoid exploitation of natural resources	cognitive	direct instruction
8	Student two	especially the marine life	cognitive	direct instruction
9	Student four	and please who cares	affective	
10	Student one	[My country's] mining industry is owned by countries like Canada, Norway, I want put forward proposals on safety measures which need to be taken to preserve the natural resources	metacognitive	instructional design

This sample of 10 statements reveals the range of learning and tutoring activities that were coded across the three chats, although the balance of activities changed markedly between and within chats. This example illustrates coding for learning activities using categories drawn from Veldhuis-Diermanse as cited by De Laat and Lally (2003, p. 27), and coding for tutoring processes using the Anderson schema cited by De Laat and Lally (2003, p. 29).

The statements coded for cognitive learning activities involved asking a content directed question (statements 2, 3, and 6); agreement or disagreement without backing (statement 4); or presenting an idea (statements 7 and 8). Three of the statements were coded as metacognitive learning activities because they involved higher order thinking in forms such as presenting an approach to carry out a task (statement 1), and elaborations of the planning choices based on reflections concerning the discussion (statements 5 and 10). Only statement 9 was coded as an affective learning activity, however, it is likely that in an analysis of an equivalent face-to-face conversation with variations in tone, posture, gesture, and expression, several of the other statements would have been understood as both affective and cognitive or metacognitive.

The coding for tutoring processes suggested a predominance of direct instruction that involved focusing the discussion on specific issues (statements 1 and 5), and presenting content and questions (statements 6, 7, and 8). Two of the statements were coded for the facilitation of discourse, since they drew in participants (statement 2) or sought to reach consensus or understanding (statement 3). Only statement 10 was coded for instructional design, since it was a macrolevel comment including a recommendation for the organisation of the conversation. Two statements could not be coded for tutoring processes.

Chat 2 was more effective, and the increase in the statements coded for direct instruction is consistent with greater focus on task communication. The small increase in "Instructional design and organisation" postings may be related to the shift of phase from agenda setting to bargaining. This phase shift introduced some uncertainty concerning appropriate bargaining procedure.

In Chat 3, the group was working most effectively to produce new proposals for subplenary and plenary meetings. The process at the level of the chats, and overall trade bargaining was generally understood, so there was a reduced need for "Instructional design" statements. Students experienced a sense of urgency to produce proposed motions, and were able to focus on the discussion of specific motions with very little overt facilitation from peers. If we consider the stages of group development, then, by Chat 3, the environment subcommittee has reached the "performing" stage.

Conclusion and Implications for CSCL/NL

In this chapter, we have looked in detail at vignettes from two case studies. We have focused on synchronous and asynchronous discussion occurring between group members in CSCL/NL, in two, quite distinct contexts, using content analysis and critical event recall. Our focus was directed to discussion by using two theoretical frameworks: sociocultural theory and constructivism. This, then, led us to focus on the context (different tutor groups [Case study 1]) and modes of discussion (synchronous and asynchronous [Case study 1 and 2]), and the nature of discussion itself, as a knowledge-building process within the groups. This was undertaken using CA (both case studies) and CER (Case study 1). In terms of our idea about Theory-Praxis conversations, this shows how theory has informed the focus of our research into praxis (albeit in general terms here). The analysis itself has provided us with rich detail about the discursive dynamics of each group. We were then able to relate this to the context within which the groups were working. We would like to argue that this approach, using these methods, is consistent with the theoretical frameworks we employed. The actual discussions analysed in these studies show us that context has a strong and specific influence upon the discursive dynamics of each group, whether synchronous or asynchronous. This aspect of praxis, as revealed by our investigation, is currently beyond the theoretical frameworks employed, in terms of their ability to direct our research focus in more detailed ways, or provide explanations of our findings. At this point then, we can see that the general theoretical frameworks need to be reshaped to reflect this. Our next task will be to look at a wider range of other contexts (in a metaanalysis) to seek some confirmation of this modifying effect of our research upon theory (work currently in progress).

As well as reporting these empirical findings and their divergence from existing theory, we would like also to argue that these case study vignettes highlight the difficulties to be faced in researching CSCL/NL. We suggest that the key research challenges to be faced by the CSCL/NL research community are

- How do we achieve a synthesis of multitheoretical perspectives, and thereby, increase their power as a tool in the research process (Halverson, 2002), and as a guide to the design of praxis?

- How do we take account of the richness and complexity of praxis, and understand the extent to which it is currently "beyond theory" in any case or context?

- How do we support the creative interaction of theory and praxis, and the research methodologies used to facilitate this?

In our present findings, the interaction between context and knowledge-building processes is not clear. It is clear that the context is shaping the nature of the group dynamics, as members strive to achieve their goals, but the precise nature of this interaction is obscure. Both theoretical perspectives have helped us, but until they are more fully integrated, their power of explanation is limited. Our studies suggest, for example, that the level of experience of the tutor is a powerful contextual factor affecting knowledge building (Case study 1), and that affective and cognitive processes may interact in powerful ways (Case study 2). These findings are beyond theory at present. The methods we have used work together to provide the richness of analysis that helps to point out these theoretical limitations to us. So, in order to take account of the richness and complexity of praxis, we need to employ multiple theoretical perspectives. In order to synthesise these, we must look at reconstructing theory in the light of new findings. To support this kind of creative theory-praxis interaction, we suggest that this approach may be employed more systematically by researchers in the field. In a recent survey of over 30 CSCL/NL papers (De Laat, Lally, &

Simons, 2005), we found that only three used their findings to discuss the impact of their work on the theoretical constructs with which they had started out. Unless we undertake this task more routinely, how can we hope to move theory forward?

In summary then, we offer these case studies as an example of how empirical analysis of discussion might be used to help us to understand CSCL/NL more deeply, and as a concrete example of the how we might begin to address challenges faced by CSCL/NL research over the next 10 years: synthesizing theory, providing richer analyses of praxis, and using both, in interaction, to develop the theory-praxis conversations in which our understandings of CSCL/NL might deepen. These findings may then be used to help build a coherent, research, evidence base upon which implementations of educational design in CSCL/NL may be built. We welcome discussion and debate about this process.

Acknowledgments

We thank students and academics of the University of Sheffield, UK, and the University of Cape Town, SA, for their help in this study.

References

Anderson, T., Rourke, L., Garrison, D. R., & Archer, W. (2001). Assessing teaching presence in a computer conference context. *Journal of Asynchronous Learning Networks, 5*(2), 1-17.

Banks, S., Goodyear, P., Hodgson, V., & McConnell, D. (2003). Introduction to the special issue on advances in research on networked learning. *Instructional Science, 31*(1-2), 1-6.

Beuschel, W. (2003). From face-to-face to virtual space. In U. Hoppe (Ed.), *Proceedings of the International Conference on Computer Support for Collaborative Learning 2003 (Designing for change in networked learning)* (pp. 229-238). Dordrecht: Kluwer.

Carr, T., Cox, G., Eden, A. & Hanslo M. (2004). From peripheral to full participation in a blended trade bargaining simulation. *British Journal of Educational Technology, 35*(2).

Chi, M. T. H. (1997). Quantifying qualitative analyses of verbal data: A practical guide. Jouranl of the Learning Sciences, 6(3), 271-315.

Cox, G., Carr, T., & Hall, M. (2004). Evaluating the use of synchronous communication in two blended courses. *Journal of Computer Assisted Learning, 20*(3), 183-194.

Damasio, A. R. (1996). *Descartes' error: Emotion, reason and the human brain.* London: Papermac.

De Laat, M., & Lally, V. (2003). Complexity, theory and praxis: Researching collaborative learning and tutoring processes in a networked learning community. *Instructional Science - Special Issue on Networked Learning, 31*(1-2), 7-39.

De Laat, M., & Lally, V. (2004). Researching collaborative learning and tutoring processes in a networked learning community. In P. Goodyear, S. Banks, V. Hodgson, & D. McConnell (Eds.), *Advances in research on networked learning* (Vol. 4, pp. 11-47). Boston: Kluwer Academic Publishers.

De Laat, M., Lally, V., & Simons, R-J. (2005). *Questing for coherence, analysing complexity: A critical synthesis of empirical findings in networked learning research in higher education* (in preparation).

Elliott, J. (2001). Making evidence-based practice educational. *British Educational Research Journal, 27*(5), 555-574.

Goodyear, P. (2001). *Online learning and teaching in the arts and humanities: Reflecting on purposes and design.* Retrieved September 1, 2002, from http://iet.open.ac.uk/research/herg/han/2001conf.htm

Goodyear, P., Avgeriou, P., Baggetun, R., Bartoluzzi, S., Retalis, S., Ronteltap, F., et al. (2004a). Towards a pattern language for networked learning. In S. Banks, P. Goodyear, V. Hodgson, C. Jones, V. Lally, D. McConnell, et al. (Eds.), *Networked learning 2004* (pp. 449-455). Lancaster: Lancaster University.

Goodyear, P., Banks, S., Hodgson, V., & McConnell, D. (Eds.). (2004b). *Advances in research on networked learning* (Vol. 4). Boston: Kluwer Academic Publishers.

Goodyear, P., De Laat, M., & Lally, V. (2005). Theory into design into theory, *European Conference on Research in Learning and Instruction (EARLI).* Cyprus.

Hakkinen, P., Jarvela, S., & Makitalo, K. (2003). Sharing perspectives in virtual interaction: Review of methods of analysis. In U. Hoppe (Ed.), *Proceedings of the International Conference on Computer Support for Collaborative Learning 2003 - Designing for change in networked learning* (pp. 395-404). Dordrecht: Kluwer.

Halverson, C. A. (2002). Activity theory and distributed cognition: Or what does CSCW need to do with theories? *Computer Supported Cooperative Work, 11*, 243-267.

Johnston, R. (1998). The university of the future: Boyer revisited. *Higher Education, 36*(3), 253-272.

Kneser, C., Pilkington, R., & Treasure-Jones, T. (2001). The tutor's role: An investigation into the power of exchange structure analysis to identify different roles in CMC seminars. *International Journal of Artificial Intelligence in Education, 12*, part II of the Special Issue on Analysing Educational Dialogue Interaction, 63-84.

Lally, V., & De Laat, M. (2002). Cracking the code: Learning to collaborate and collaborating to learn in a networked environment. In G. Stahl (Ed.), *Computer support for collaborative learning* (pp. 160-168). Hillsdale, NJ: Lawrence Erlbaum.

Lally, V., & De Laat, M. (2003). A quartet in e: Investigating collaborative learning and tutoring as knowledge creation processes. In U. Hoppe, B. Wasson & S. Ludvigsen (Eds.), *Designing for change in networked learning environments* (pp. 47-56). Amsterdam, NL: Kluwer Academic Publishers.

Ontario Institute for Studies in Education. (1997). *Computer support for collaborative learning '97*. Retrieved November 15, 2004, from http://www.oise.utoronto.ca/cscl/

Paavola, S., Lipponen, L., & Hakkarainen, K. (2002). Epistemological foundations for CSCL: A comparison of three models of innovative knowledge communities. In G. Stahl (Ed.), *Computer support for collaborative learning: Foundations for a CSCL community* (pp. 24-32). NJ: Lawrence Erlbaum Associates.

Stahl, G. (Ed.). (2002). *Computer support for collaborative learning: Foundations for a CSCL community*. Hillsdale, NJ: Lawrence Erlbaum.

Stahl, G. (2003). Building collaborative knowing: Contributions to a social theory of cscl. In W. Strijbos, P. Kirschner, & R. Martens (Eds.), *What we know about CSCL in higher education* (pp. 53-85). Dordrecht: Kluwer.

Taylor, C. (1991). *The ethics of authenticity*. Cambridge, MA: Harvard University Press

Veldhuis-Diermanie, A. E. (2002). CSC *Learning? Participation, learning activities, and knowledge construction in computer-supported collaborative learning in higher education*. Wageningen: Grafisch Service Centrum Van Gils.

Wasson, B., Ludvigsen, S., & Hoppe, U. (2003). Designing for change in networked learning environments. In U. Hoppe (Ed.), *Proceedings of the International Conference on Computer Support for Collaborative Learning 2003 - Designing for change in networked learning* (pp. xvii-xx). Dordrecht: Kluwer.

Endnote

[1]   By CSCL/NL we mean education that uses Internet-based information and communication technologies to promote collaborative and cooperative engagement: between one learner and other learners; between learners and tutors; between a learning community and its learning resources, so that participants can extend and develop their understanding and capabilities in ways that are important to them, and over which they have significant control (Banks, Goodyear, Hodgson, & McConnell, 2003).

Chapter X

Identifying an Appropriate, Pedagogical, Networked Architecture for Online Learning Communities

Elsebeth Korsgaard Sorensen, Aalborg University, Denmark

Daithí Ó Murchú, Gaelscoil Ó Doghair &
Innovative e-Learning/e-Tutoring, Hibernia College, Ireland

Abstract

This chapter addresses the problem of enhancing quality of online learning processes through pedagogic design. Based on our earlier research findings from analysis of two comparable online master courses offered in two masters' programmes, respectively from Denmark and Ireland (Ó Murchú & Sorensen, 2004), we present what we assert to be a fruitful, student-centred, pedagogic model for design of networked learning. The design model is composed of what we have identified as unique characteristics of online learning architectures that, in principle, promote and allow for global intercultural processes of meaningful learning through collaborative knowledge building in online communities of practice. Inspired by principles of best practice and Wengerian design criteria for networked learning (Wenger, 1998), the chapter intends to access, discuss, and provide evidence associated with the quality issues of the presented model, and its specific learning architectures.

Introduction

All the world's a stage
And all the men and women merely players:
They have their exits and their entrances;
And one man in his time plays many parts.

(Shakespeare)

The overall educational challenge of today's knowledge society is the enhancement of learning through ICT. It requires creativity, as well as innovation and change in educational thinking to identify educational paradigms that allow for learning processes to unfold in networked contexts, in ways that truly provide quality in processes of learning (Bates, 1999; Collis, 1996; Collis & Moonen, 2001; Miyake & Koschmann, 2002). Consequently, the complex task of generating pedagogic learning architectures conducive to learning in networked environments is becoming increasingly more important (Sorensen, 2004).

However, a high proportion of stand-alone-like educational designs suggests that the necessary innovation and change in educational thinking appear to be the exception, rather than the rule. Many educational designs unfolding in networked contexts mirror the inherent assumption that learning does not involve inter- and intrapersonal human interaction and, "legitimized" by impeding economic considerations, that the task of designing for learning is solely a matter of prediction, formalization, and preparation of software-instructional processes to be used between learner and learning software.

In contrast, extensive research (Bates, 1999; Collins, Mulholland, & Watt, 2001; Collis, 2001; Harasim, 1999; Harasim, Hiltz, Teles, & Turoff, 1995; Koschmann, 1996) indicates the significance, in terms of learning quality and outcome, related to the social aspects of a learning process. Furthermore, it is widely accepted that the potential of network technologies lies in their facilitation of interhuman interaction (Harasim et al., 1995; Mason, 1993; Mason, 1998; Sorensen, 1993; Sorensen, 2004; Sorensen & Takle, 2002). This potential, the true soul of technology (Ó Murchu, 2005), should be sought in their ability — independent of time and space-to enable and facilitate tapestries of networked learning designs, which, emphasized by their underlying philosophy, aim to promote the advancement of non-authoritarian, democratic, student-motivated, and -directed learning designs, resting fundamentally on collaborative knowledge building processes. Thus, in essence, the core educational principles of such networks are to promote student activity in terms of facilitating collaborative knowledge building among participants in a shared endeavour, across time and space (Salmon, 2000; Scardamalia & Bereiter, 1996; Sorensen & Takle, 2004). It seems extraordinary how modestly this collaborative knowledge building potential is generally utilized in networked learning designs, and how plentiful the myriad of online learning designs that continue to bring into the virtual arena various methodologies directly transferred from face-to-face teaching/learning paradigms.

Taking these research findings into account, the challenge of creating pedagogic-didactic designs of good quality must inevitably depart from a design perspective on learning as a social matter, and aim at establishing learning designs which allow for processes of student collaboration that unfold spontaneously, in unpredicted, democratic, meaningful, and encultured tapestries of responsible knowledge building dialogues (Ó Murchú, 2005).

In an overall perspective, the "design" challenge may be envisioned, using the metaphor of "space," and directed towards creating "learning architectures" (Wenger, 1998; Wenger, McDermott, & Snyder, 2002), or evolving membranes of design quality. More precisely, acknowledging the fact that it is not possible to directly "design learning," but only to "design for learning" (Wenger, 1998), the initial design focus should be directed towards the features of such learning architectures or membranes. This means addressi-ng the *multi-dimensions* of the learning space, which invite and incite the multiple intelligences of the participants to experience "soulful" learning as defined by PLATO (Ó Murchú, 2005).

This chapter addresses the problem of enhancing quality of networked learning architectures through pedagogic design. For this purpose, we suggest what we believe to be a fruitful pedagogical framework for such design endeavour, and one that supports the learning values we wish to promote. Inspired by various theoretical perspectives on learning and on the pedagogic design, and with reference to earlier findings (Sorensen & Ó Murchú, 2004), we describe, discuss, and demonstrate the specific qualities of the proposed learning architecture.

Background and Theoretical Perspective

We can identify two, distinctly different approaches to design of networked learning (also commonly referred to as "e-learning"). Viewed on a large scale, the first approach concerns the vast majority of e-learning designs. These tend to aim at facilitating individualized learning processes (Mason, 1998), and are commonly based on the use of computer-based training (CBT) and multimedia applications. The methods generally used to facilitate and support such models of e-learning rest on what we could call "stand-alone concepts" involving, in broad terms, the notion of learning as a phenomenon, which takes place individually and cognitively in a mental space without social impediments. Basing learning on such a notion and, consequently, applying methods for delivery without social inputs, have a strong impact on the nature of the design task and challenge. The quality of design in such cases will depend on the ability of the designer to support the pedagogic-didactic methods of the learning process through *foresight, prediction*, and *formalization* (Sorensen, 1993). This type of e-learning tends to be regarded, especially by companies, as an economically feasible solution. The costs are more easily controlled, as they are mainly related to the process of design, development, and planning: whereas the delivery process, generally involving only the user and the design, is very low cost.

The second approach appears to have a different point of departure. From a human educational perspective concerned with the production of human competencies needed for engaging in society as a democratic and critical citizen, the standing approach to development of e-learning — if it were to spread more generally to educational institutions — gives rise to some concern. Nevertheless, even if competition is hard, more interhuman based design concepts have developed over the years, most frequently within contexts of educational institutions. In fact, we may speak about "principles of best practices of e-learning" (Ó Murchú, 2003) or online education, designed from approaches that to a smaller or larger extent, involve interhuman processes.

Regardless of the differences in these two ways of interpreting the challenge of design, it is a fact that the core potential of network technologies lies in their facilitation of interhuman interaction (Harasim et al., 1995; Sorensen, 1993). In their ability — independently of time and space — to enable and facilitate learning designs that, as their underlying theoretical philosophy, aim to promote the advancement of non-authoritarian, democratic, and student-centred learning designs. In essence, these networks possess the potential for promoting students' activity (Sorensen, 2004) in terms of:

- facilitating an equal interaction and collaboration among participants (collaborative knowledge-building dialogue in a shared endeavour, across time and space);
- enhancing reflection on action;
- breaking down power structures; and
- promoting democratic processes of negotiation.

Therefore, it seems striking how infrequently this potential is utilized in online educational designs. Many designs still bring onto the virtual stage, methodologies specifically inherited from face-to-face teaching/learning paradigms. Moreover, they fail to stimulate change and birth of innovative, didactic features and methodologies that, in more appropriated ways, utilize this potential on democratic and student-centred grounds, and thus, induce change in the roles of students and teachers.

"Design"

Design of online learning has continuously challenged the educational world for more than two decades. The term "design" frequently causes confusion. On the one hand, it tends to be associated either with technical features and structure of the online learning environment, or with the composition of the human-computer interaction aspects of using technology. On the other hand, it is associated with the branch of knowledge concerned with research and theory about instructional strategies, as well as the process of developing and implementing those strategies. In this category, also known as "instructional design" (Wilson, 1997), it is viewed as the entire process of analysis of learning needs and goals, and the development of a delivery system to meet those needs.

Thus, the "design" concept has many connotations. Salomon (1992) referred to music, using the metaphor "orchestrating," to describe and bring about new aspects of the challenge of "design" of technology-driven learning environments. We, on the other hand, are moving towards a dramatic metaphor, and invite the reader to perceive networked learning as a "play of learning," and see the challenge of pedagogically and didactically designing networked learning as a matter of "staging" such a play. In doing so — and from the point of view that "learning happens, design or no design" (Wenger, 1998, p. 225) — we acknowledge the inescapable fact that the perspective we, as designers, bring along ought to be a conscious one, since our design decisions and choices in the process of "scripting," shape what we envision and identify as essential pedagogical tasks to be addressed in a given situation.

Design Ethos

An attractive vision of learning addressing, in particular, the individual level of learning, is proposed by Paul Colaizzi (1978) as representative of an existential view on learning. His concept of "genuine learning" may be generally described as including the following characteristics:

- It is rare;
- It is difficult;
- It is individual, but stimulated collaboratively;
- It is situationally unpredictable;
- Learning programmes do not automatically produce genuine learning, but can, at the most, inspire it;
- It changes our views and perceptions of the world;
- It has an extension in time and can never be fully finished;
- It gives an insight that leaves no doubt;
- It creates existential commitment (with an element of risk) as it has to do with the meaning of life;
- It chooses existentially, and with an attitude of respect;
- It is patient;
- It is authentic learning;
- It is the unlearning of bullshit; and
- It is never dull.

Learning in Online Communities of Practice (COPs)

Wenger (1998) presents the idea that "design" is a question of creating learning architectures of good quality. Assuming that it is not directly possible to "design learning," he argues that the initial design focus should be directed towards the features of the learning architecture. This means addressing, initially, the *dimensions* of the design space, among which the design decisions concerning the learning context are placed (Sorensen & Ó Murchu, 2004). These dualities constitute the basic principles of a learning theoretical perspective, viewing learning as a matter of *interaction between experience and competence*. They frame and define the pedagogic "space" and its qualities for enhancing such a learning perspective, and they constitute the "walls" of the learning space within which a pedagogic design, and its various implied techniques, can be decided upon. Given such multi-dimensional learning space as a prerequisite to the more detailed implementation of pedagogic techniques, the *motivation* (the drivers of the learning process) needs to be considered and facilitated.

"Best Practice"

The concept of "best practice" developed from a movement by curriculum speciality organizations such as the National Council of Teachers of English, the National Council of Teachers of Mathematics, the International Reading Association, and many others emphasizing standards: it was originally developed by Daniels and Bizar (1998). Although initially addressing traditional teaching and learning situations, the concept has been adapted, reconstructed, and further extended by Ó Murchú (2003) to fit with networked learning environments. In terms of design for best practice, "practice" so far includes certain design guidelines for educational activities, techniques, and methods for learning, teaching, mentoring, and assessing (Daniels & Bizar, 1998; Ó Murchú, 2003).

The two approaches, learning through online knowledge building in COPs, and learning from experiences of good practice, represent an interesting analytical combination. Together, they complement each other, and form a balanced optic for our discussion on features of appropriate learning architectures, as they are derived from two opposite "areas of knowledge": Wenger's model (1998), representing theoretical knowledge and his suggestion for the fundamental and "existentially" flavoured drivers of the learning process; and Daniels & Bizar (1998) and Ó Murchú (2003), in their framework, concentrating on the more "experiential" and tangible pedagogy and its techniques developed over time within the area of practice.

Research and Design Methodology

The empirical context for our meta-analysis is constituted by the findings from earlier analyses and investigations of two, comparable online courses from two Master programmes designed in Denmark and Ireland, respectively (Ó Murchú & Sorensen, 2004):

Course on "online learning" (OL)[1]

The course on "online learning" is one of several courses on the Danish cross-institutional online MS in ICT and Learning (MIL). MIL is a two-year (part-time) Master education in ICT and Learning. MIL provides continuing education for working adults engaged in educational planning and integration of ICT in learning processes at schools and all types of educational institutions. Employees with educational responsibilities in different types of organizations also enter the program. MIL is structured in four categories of studies: four modules (each consisting of three to four courses), one project work, and one Master thesis. Many of the approximately 40 MIL participants were highly qualified teachers at the high school level. They have extensive university education and high competence within their individual work areas.

Course on qualitative research methodologies and online learning possibilities (MIC)[2]

This Masters course (MEd. & MA in Education) is a two-year taught programme that is comprised of working adults from various walks-of-life in education (elementary, secondary and third level), adult education, private sector, and business Similar to MIL, Denmark, the MIC programme is structured in four categories of studies: four modules (each consisting of three to four courses), one project work, and one Masters thesis. The majority of participants (56%) were practicing teachers, or involved in education management (18.6%). The remainder hailed from administrative or private business and general education (25.4%). They had extensive university education, and high competence within their individual work areas. Having firstly completed a preliminary Diploma at the University, and submitted a "project" on a topic of their own choosing, the participants were deemed eligible to further their studies to Masters level in the second year by attaining a recognized "honours" standard in their work.

The tools used in both courses to collect the data for our investigation and analysis of the formation of online communities of practice (COPs), were questionnaires developed to produce both quantitative insight on the development and features of COPs, and qualitative statements from participants in the course delivery process.

Our earlier findings (Ó Murchú & Sorensen, 2004; Sorensen & Ó Murchú, 2004) tentatively suggest that both courses, despite diversity of the backgrounds of the two course designs, produced the forming of COPs. We also asserted that processes of learning unfolding in learning environments, or "learning architectures," featuring characteristics of COPs (as in the case of the two courses) enhanced networked learning. One specific element in the responses to the questionnaires that caught our attention was the frequent statements related to the significance of the role of the teacher.

The conclusion of our study, that the design philosophies of both courses were captured and "accepted" in a Wengerian optic (Ó Murchú & Sorensen, 2004: Sorensen & Ó Murchú, 2004), produced the hypothesis, which we explore in this chapter:

That a more effective design model for enhancing learning through promoting the formation of online COPs may be achieved by merging the features of the two design philosophies into one design architecture and placing a stronger emphasis on the changed role of the teacher/designer.

In stating this hypothesis, we assume that the design features, resulting from the two design philosophies and designs of the master courses investigated, are essential for the forming of online COPs.

In this study, we employ the perspective of the teacher/designer. We meta-analyze, discuss and illustrate, theoretically and through comments from participants, the features of our newly born design architecture, the Model of Merged Design (MMD). The discussion is inspired by the rhetoric of drama, and draws on the theoretical perspectives introduced above, as well as elements and key issues identified by the authors through many years of practice as researchers and designers within the field of ICT and learning.

Capturing the Walls of the Learning Architecture: The MMD Model

The ambitions and visions that led to the design of the Danish OL course are, to a large extent, a result of the insights obtained through 15 years of practice and research within technology-supported open and distance learning, and the influence from the Danish pedagogic tradition, and its historical preferences of "the living dialogue" (Sorensen, 1997). In contrast, the ambitions and visions behind the qualitative research methodology module of the Irish MIC course were, to a large extent, determined by tangible pragmatic conditions. As such, its design represents a contrasting alternative to the traditional, and was confined within the possible choices of action of the specific political, cultural, and geographical context.

Viewed in a meta-analytical perspective, the two course designs and delivery processes mirror, to a smaller or larger degree, preceding design considerations as included in the following list (Sorensen & Ó Murchu, 2004):

- Which elements to structure and make procedures for on the basis of prediction?
- To what extent the design should depend on decontextualized knowledge?
- How to balance student initiative/ownership and pedagogical authority?
- How to minimize teaching (the predicted) in order to maximize learning?
- How to maximize processes of negotiation of meaning enabled by interaction?
- How to broaden the scope of coverage without loosing the depth of local engagement?
- To whom, and in which way does the design represent an opportunity to build an identity of participation?

Pooling the above two design philosophies, which both, apparently, produced pedagogical features and techniques that stimulated the development of online COPs (Ó Murchu & Sorensen, 2004), and their lists of design considerations, we may end up with a model (MMD) (inspired by Sorensen & Takle, 2003), as shown in Figure 1.

Combining the design features of the two courses into one learning stage (the MMD model), we expect to constitute an even stronger learning architecture for potentially stimulating the kind of "scripts" for performing networked "plays of learning" that will enhance networked learning through the formation of true online COPs.

Figure 1. Model of Merged Design (MMD)

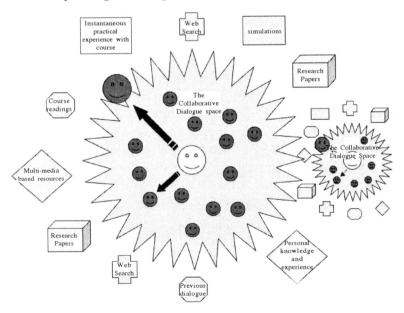

A Collaborative Dialogue Space

The collaborative dialogue space is the structuring centre of the teaching/ learning process: the basic feature of the learning architecture offered by the MMD model.

(...) learning is, in its essence, a fundamentally social phenomenon, reflecting our own deeply social nature as human beings capable of knowing. (Wenger, 1998, p. 3)

(...) it is in the meanings we are able to negotiate through learning that we invest ourselves, and it is those meanings that are the source of the energy required for learning. (Wenger 1998, p. 266)

The collaborative dialogue space is where "the play of learning" comes into existence through a tapestry of dialogue. It is where the strongest collaborative energy of a learning group manifests itself in the "Now," the instant of shared dynamic dialogue and strongest energy between participants:

Nothing ever happened in the past; it happened in the Now. Nothing will ever happen in the future; it will happen in the Now. What you think of as the past is a memory trace, stored in the mind, of a former Now. When you remember the past, you reactivate a memory trace — and you do so now. The future is an imagined Now, a projection of the mind. When the future comes, it comes as the Now. When you think about the future, you do it now. Past and future obviously have no reality of their own. Just as the moon has no light of its own, but can only reflect the light of the sun, so are past and future only pale reflections of the light, power, and reality of the eternal present. Their reality is "borrowed" from the Now. (Tolle, 1999, pp. 41-42)

The dialogic "Now" is the melting pot for the forming of a shared learning endeavour between participants, opening up the possibility in the learning space for the unfolding of an actual democratic interaction between learners, in which each perspective thrown into the discussion, in principle, has the potential of pushing the learning process, individually and collectively, through collaborative knowledge building (CKB), and of initiating change. Examples of student perceptions and comments illustrating the power of the CKB process are:

I felt like I was part of everyone's learning, like a family of thinkers on the net. It was a great feeling of confidence. (Student comment)

There was a constant active exchange of ideas and thoughts. Never a dull moment for the grey cells to rest. (Student comment)

The learning process induced by the MMD learning architecture is *democratic* in the sense that it evolves, not according to authoritative powers (teacher authority), but according to the democratic principle of equality (everybody has a voice). It is the quality of the argument — the competence of the speaker — that manifests itself most profoundly on the road to change.

The collaborative dialogue space of the MMD model represents the scene of the learning process characterized by shared and negotiated knowledge building between participants in the *power of the Now*. The play of learning may unfold according to a higher or lower degree of prediction and formalization (as responses to tighter or looser "scripting", allowing for improvisation of the actors) depending on the actual learning context, and its specific conditions with respect to organisation, target group, and so forth. The power of the Now also extends to the dimension of *social presence*, as illustrated by a student comment:

I always felt that it was up to me to set the goals and direct my own learning but I never felt alone in my work. The group was always there for me, anytime any place, the joy of exchange and thinking together. (Student comment)

The interactive space is where the expectation to meet and interact with others appears and functions as a vital motivator for the general engagement, and for the vision to authentically, influence and make a difference in the knowledge building process. The sense of openness of the model in terms of, on the one hand, unpredicted initiative and individual freedom to put a comment in the dialogue space and, on the other hand, the collaborative sense of co-presence in that same space, seems an essential driver of the learning process, utilizing a very ontological, interhuman mechanism of interhuman existence. This is a very strong feature of a networked learning process and of the MMD model.

A Wide Resource Concept

The MMD architecture operates with a multimodal and very wide and diverse concept of resources. These may be of any kind of nature, ranging from traditional literature and readings of research papers, pieces of software, personal/mutual experience, and expert knowledge, to "meta-resources" like, for example, previous dialogue and other plays of learning (Figure 1). This wide resource concept adds to the openness of the model. Minimizing the determination of the script (the predicted) of the play of learning, it leaves the actors with a freedom to establish ownership, to improvise, and thereby influence in self-motivated meaningfulness, the CKB process. In principle, any type of resource that enhances the CKB process may be pulled into the discussion by the participants.

The open and diverse resource concept of the MMD architecture allows a continuum of possibilities in terms of being dependent, or not being dependent, on decontextualized knowledge; a balance between narrow instructional structure and pedagogical authority (tight instructional scripting) that may lead to students reproducing; and an open stage where the teacher steps aside and leaves students to construct knowledge (loose instructional scripting), influence their learning process, and take ownership, to broaden the scope of coverage, without loosing the depth of local engagement (the need to be detached from practice vs. the need to be connected to it).

The MMD architecture offering and employing the energy of the Now in the CKB process, denies that dialogue and negotiation processes leading to learning are of a reproductive, and potentially alienating nature. Elements of requests of "reproduction" appear in the MMD learning stage, not in the process of the Now, but belonging to the concept of "resources" in the shape of smaller elements of training. Processes of "reproduction" assuming a view on learning as "transfer" are needed in various learning situations, but only as resource elements in a larger learning process, tied together, with the energy of the Now, in a CKB process. The MMD learning stage, with its central emphasis of the CKB play as the ultimate learning process, invites new forms of identity building and negotiability. It is based fundamentally on the principle of legitimate peripheral participation (Lave & Wenger, 1991), in which participants are expected to contribute their own views, knowledge, and competence to the CKB process, creating meaningful forms of membership, and empowering forms of ownership of meaning. This also means opening up for new trajectories of participation. As a consequence, new opportunities and scopes for building identities of participation become possible, continuously, throughout the CKB process of negotiation.

Reflection, Meta-Learning and Identity

The MMD architecture is relying heavily on the dimension of reflection and meta-reflection: a dimension that is vital, not only for the building of identity. It is also, according to Bateson (1976), an essential element in the very constitution of a learning process. Bateson views all learning as taking place in the shape of reflective movements between meta-communicative levels.

The reflective and meta-reflective aspects of the MMD model manifest themselves in a variety of ways. A prime emphasis on collaboration and dialogue, and on the dialogic process and methodology in learning, stimulates a style of learning that implies a meta-reflective practice of democratic methods and techniques of negotiation. A student commented as follows:

I was always thinking beyond the written facts as my friends and I exchanged ideas and responses. Everyone's responses were treated equally and with respect and at all times I was challenged to participate and reflect not just as a student but as a person of importance, whose ideas mattered

(Student comment)

More precisely, this is implemented in the networked learning architecture using a structure of fora for the collaborative dialogue, a meta-fora structure for the meta-discussions of the collaborative dialogue, and so forth. Below is an exchange of student comments which illustrate such meta-reflective communicative practice unfolding in a meta-forum for the collaborative dialogue space:

Just up to breathe for air: I think, it is frustrating not to be able in any way to relate qualitatively to all of these exiting comments, which flow in a constant stream in the fora (and I am only a member of the ones that are relevant to me). The question is if it is a "learning frustration"? E.g.. "learning to prioritise"? Or if this is more likely a case of information overflow? In other words, I am wondering whether the amount of 50 active students is too high for this type of fora? Has there, by the way, been any research carried out on this topic — I mean what is too much and what is too little in terms of the appropriate amount of students in asynchronous, distributed learning fora?....I'm diving again now, I am breathing well again now :-)

(Student A)

Hi <Student A>,

Well, perhaps it is about developing some kind of overview of the various threads that are being discussed, and then selecting those that one finds interesting to comment on (and then leave the rest, browse them, and then that's that). Then you can comment on those which you find exciting or

which awaken something in you...or you can start a new thread, if you have
something on your mind that has not been touched upon....I know that it
sounds very "wise"...but it is probably a kind of survival strategy that you
have to come up with. If we learn how to behave in a society of information,
where we are constantly bombarded with all kinds of information, then we
may be able to teach others...and that, I guess is something we need to learn
also (sorting, I mean)....But yes, you have a point....there is so much to
chose between.....

<Student B>

Hi <Student B>,

I agree completely. That's what I'm trying to do. But the difference between
MIL and the information society is — for me — that I feel much more
committed to MIL. In other words, I don't find it terribly exciting. I really
don't want to miss out on something exciting :-)....And after all, my mood is
still very high! :-)

<Student A>

The ethos underlying the model contains a wish to stimulate learning processes, and produce global citizens that are able to further, practice, and enhance collaborative learning across diversities of different kinds (geographical, cultural, political. etc.). These are, typically, learning processes that are based on non-authoritarian and democratic values, where a critical listening to the opinion of others in taking a stand is a vital meta-learning element, also referred to as "deutero-learning" (Bateson, 1976), denoting learning about how to learn (i.e., learning about one's self and learning). The online collaborative, dialogic request and emphasis of the MMD model stimulates such reflective meta-learning, or self-inspection and awareness at different reflective levels, on the basis of the characteristic dimension of distance, which is an inherent valuable characteristic of a networked stage, inviting a duality of dialogic and reflective thinking independently of time and space (Sorensen, 2004). This is a vital feature of the MMD model, as educational reification somehow inserts what could be named an extra artificial level, or stage, between practices and learners.

The reflective meta-learning principle, as different levels of awareness (Bateson, 1976), is indicated in more detail in Figure 2. Sorensen (2004) argues that entering a virtual networked reality enhances reflection by excluding the first "unconscious" level of learning. When entering the virtual space, the learner arrives at level 2 as the minimum possibility (Figure 2).

In the MMD walls of the learning architecture, this principle is indicated in the right side of the model, denoting the collaborative dialogue space, itself understood as a resource for reflection (Figure 1), employed in parallel to its process of evolvement, as well as in retrospection of a reificative product of the learning process. We may understand these reflective processes and movements between different communicative levels, in line with the dramatic technique of "Verfremdung," introduced by the German dramatist, Bertolt Brecht (1968). His ambition was to create reflective spaces that involve the audience, the learners. Brecht's concept of "Verfremdung" may be seen as a meta-reflective part of a learning process, or a process of building awareness at various meta-levels.

The enhanced *visibility* to and *transparency* of the scripting of the learning process also invites (through its visible mediated manifestation) invention and self-motivated participation (in the scripting of the learning play) from participants, through the mediated manifestation of both *individual* and *collaborative* desires and practices of the dialogic process. Examples of student comments that demonstrate this dual motivation are

I really believe that this course has shown me the importance of sharing everyone's ideas equally as shared-reflection leads to deeper understanding.
<Student comment>

There was never a night that did not feel good and so interesting. I could surf away and yet feel free to share and reflect. No one drove me to learn. I directed myself and felt so much a part of something wonderful and exciting. I heard, I saw, I listened, I shared, I thought, I reflected, I collaborated and I was equal in the online family. Daithí called it a community of practice. I call it a family of trusting thinkers, not afraid to think out loud or in silence but always welcome to be listened to.
<Student comment>

Figure 2. Enhanced meta-learning condition in networked learning (Sorensen, 2004)

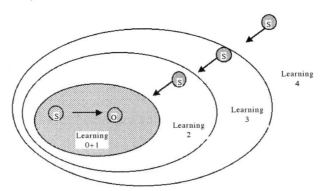

Teacher-Student Roles

Learning, with or without teachers, takes place everywhere and at all times. Meaningful learning must be centred, motivated, and give meaning to the individual learner at the existential level, if he or she is to truly learn, especially from the perspective of a meta-learning evolving towards the level of genuine learning.

The learning architecture of the MMD model implies a supple and dynamic approach in terms of designing/scripting teacher-students roles. It works from the principle that the greater the emphasis on participants' experiences, ownership, authenticity of true motivation, genuineness, non-authoritarian teaching/learning, and so forth, in an envisioned pedagogical design, the more the teacher moves away from being the deliverer of content (storyteller), and the most important and centrally placed actor in the play to becoming (1) a peripheral meta-figure preparing the stage for its play of learning from the edges of the learning process, and (2) a participant, like others, in the learning process and project (Figure 2). In the same way, as the learners become the central players, the teacher moves away from the centre, and leaves the collaborative learning stage to the learners to take ownership and initiative in what, to each of them, will represent as an authentic and meaningful learning endeavour *enroute* to an existential, genuine identity. The following comments from students indicate the time of such change of teacher roles.

I was afraid at first of being without a teacher in a classroom but realized that I got so much more as I also met many teachers and fellow students in an equal classroom
<Student comment>

I now know that my perceived role as a teacher — the centre of the classroom- is not really feasible anymore as the kids I teach are also teachers in their own right and should be encouraged to think for themselves and innovate.
<Student comment>

Moving out of the centre of the learning process, the role of the teacher changes and becomes more of a designer/script writer who delivers the pedagogical architecture for the "learning play" before it starts, and then, throughout the play, acts as a participant, learning and facilitating in the networked COPs through the various "movements between meta-communicative levels" of the networked

CKB dialogue. Wilson (1997) describes needed change in the teacher role as follows:

The role of the designer is then to design a series of experiences-interactions or environments or products intended to help students learn effectively. Neither the instruction nor the assessment of learning can be as confidently dictated as thought to be possible in the past. But the important thing to keep in mind is that the design role is not lost in such a revised system; the design still happens, only it's less analytical, more holistic (...). Instruction thus becomes more integrally connected to the context and the surrounding culture. (Wilson, 1997, n.p.)

While the MMD architecture assumes and induces such needed change in the teacher role, it is dynamic and elastic, and captures a variety of manifestations of teacher profiles. Some identified teacher role tendencies, captured by the model, are:

- **Instructional Designer:** An instructional designer considers the available resources to produce well-designed activities for student needs (Kozma, 1994).

- **Trainer:** Trainers give individual instruction to enable skilled development (in relation to the MMD model, these are events taking place within a resource level). Such training or mentoring is typically accomplished through modelling use of multimedia and technology, and helping students see how they might use software tools to accomplish unique learning tasks.

- **Collaborator:** A collaborator denotes various activities that teachers may undertake to work with their colleagues for improvement of instruction (Kozma & Schank, 1998). These activities include informal sharing with colleagues, and team teaching. They also include collaborating, sharing, and learning with the students as equals, "I am convinced that the best learning takes place when the learner takes charge"(Papert, 1993, p. 25).

- **Team Coordinator:** Team coordinator is the opposite of team manager. While team manager implies an above to below authoritarian style, a team coordinator depicts a shared vision of ownership of the learning process.

- **Enabling Advisor:** A common term used sometimes to describe this role is the term "facilitator." The MMD architecture is "membraned," spiral and ever evolving and the traditional term "teacher" becomes "enabler."

In any case, moving from traditional learning settings to a networked learning architecture constructed on the basis of the MMD model opens up, not only for the changes in teacher profiles, but typically, also, for the following ambitions and changes in the teaching/learning process:

- From teacher-centred instruction to democratically-oriented instruction;
- From single sense stimulation to multisensory stimulation;
- From single path to multi-path progression;
- From single media to multimedia;
- From isolated work to collaborative work;
- From information delivery to information exchange;
- From passive learning to active/exploratory/inquiry-based learning;
- From actual, knowledge-based "reproduction" to critical thinking and informed decision-making; and
- From reactive response to proactive/planned action.

In summary, the teacher (understood as the designer) holds the key to innovation. His/her envisioned self-perception seems one crucial consideration in the design of an online learning architecture. The MMD model envisions the teacher/ designer "on the side" of the CKB process, and supports a non-authoritarian teaching methodology aiming at equality between teachers and students, in terms of knowledge power relationships. Finally, the MMD model, through its fundamental emphasis on communicative behaviour, provides a learning architecture that may be constructed to also capture teacher profiles that address the problem of absent intentionality (Kirkeby, 1994) in networked learners, through practicing the pedagogical technique assuming what Colin Wilson has termed "the Tom Sawyer effect" (Wilson, 1998). Tackling the problem of absence and lacking communicative participation, perception, and commitment in collaborative and dialogically based processes in online learning networks, is a frequently cited problem of great complexity (Bates, 1999; Collis, 2001; Harasim, 1999; Harasim et al., 1995; Kaye, 1989; Koschmann, 1996; Mason, 1993; Mason, 1998; Salmon, 2000; Sorensen, 1993) that may be approached by a teacher attitude and profile providing the good example in acknowledging the engaging mechanism of "mimetic desire," and of linguistic intentionality, to be enacted in the energetic power of the Now, in the collaborative online dialogue space:

But the real moral, as Twain recognizes, is that "in order to make a man or a boy covet a thing, it is only necessary to make the thing difficult to attain." In other words, anything on which we concentrate our full attention becomes desirable. And anything that is difficult to attain becomes twice as desirable. When I perceive a thing, I throw my attention at it like a spear, and if I look at something idly and absent-mindedly, I often fail to see it because I am not paying attention. (...) Life is full of things that we assume to be boring and unpleasant, and which are really no more boring than things we pay money for. Everything depends on our attitude. Life is not boring or exciting; it is we who are bored or excited. (Wilson, 1998, pp. 16-17)

Meaningful and Soulful Learning

To the ancient Greeks, the root word for "soul" is the same as the word for "alive," and to them, the soul was what made living things alive. Plato considered the soul to be the "essence" of a person that reasons, decides, and acts. He considered the soul to be a separate entity from the living body, and to be immortal. In early Hebrew thought, "soul" represented the life force. However, over time, it began to be seen as something independent of the physical being. According to the Hebrew bible, when God created Adam, he "breathed" into his nostrils the breath of life; and man became a living soul. The Hebrew word for "breath" is often used to mean "spirit" and "inspiration." "Soulful learning" is therefore defined as being the essence of breathing life into transformative reflection, which comes from the inside-out.

Learning for learning's sake isn't enough....We may learn things that constrict our vision and warp our judgment. What we must reach for is a conception of perpetual self-discovery, perpetual reshaping to realize one's goals, to realize one's best self, to be the person one could be. (Gardner, 1983, p.13)

The concept of "meaningful learning" may be defined from several perspectives. From the theoretical perspective of existential learning, as presented by Colaizzi (1978), "meaningful learning" is, first of all, authentic learning. In Figure 3, we further describe and expand upon the related attributes of "meaningful learning" (Bhattacharya, 2002).

Meaningful, soulful learning, a central characteristic of the play envisioned to unfold at the stage of a MMD-networked learning architecture, is envisioned to mobilize, in the learner, the latent divinity that has always existed indigenous to the human race. The ultimate goal of any learning design is to bring out this potential as the authentically motivating force in complex learning trajectories, characterized by freedom to explore, and by self-directed and responsible learning actions leading to transformative reflectivity.

Transparency vs. Obscurity

The MMD architecture advocates transparency and participant involvement in design of learning process and product. It stimulates a dialogic, reflective, and mutually shared play of learning between learners and teachers/designers, as actors in the play. The roles taken at a given point in time during this process are the ones needed at that point to make the play of learning progress according to its design/script. Design participation and decision-making based on elements of obscurity, and manipulation alienating to the learners are not envisioned in the MMD model. The model is open for true democratic learner participation, that is, open for learners to enter, influence, and decide upon the design level by being part of processes of both the constituting scripting and acting in the learning play. Phrased in the words of Wenger (1998):

As a consequence [of the duality between the local and the global] design for learning cannot clearly separate between conception and realization, between planning and implementation; it must instead aim to combine different kinds of knowledgability so they inform each other. A design,

Figure 3. Attributes of meaningful, soulful learning (Bhattacharya, 2002)

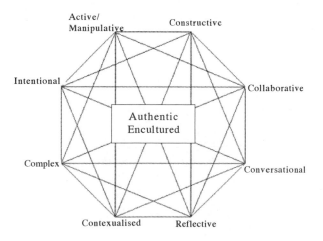

then, is not primarily a specification (or even an underspecification) but a boundary object that functions as a communication artefact around which communities of practice can negotiate their contribution, their position, and their alignment. (Wenger, 1998, pp. 234-235)

Global Learning

The MMD networked design model provides a networked learning architecture that is easily adapted for a global learning stage. In fact, a global application would utilize the model's inbuilt condition of diversity among participants and cultures even more. Tony Kaye provides a poetic wording of the ying-yang phenomenon of "learning together apart" (Kaye, 1993).

If globally employed, the MMD model, by transcending many gaps, may possess the potential of a tool promoting the "building of bridges" across not only geographical distances, but also cultural and political diversity through nurturing the establishment of fruitful global online networks of learning COPs.

Conclusion and Future Perspectives

Our aim for this chapter was to construct a generalized, dialogic learning architecture for design of networked online collaborative learning for adult learners. We constructed the MMD model through applying a meta-optic on our previous research findings from two comparable online courses from two different master programmes, analysing to what extent the two courses had developed into networked communities of practice. Our findings showed the evolution and existence of COPs. Working from the hypothesis that an enhanced and empowered model for design of networked, learning COPs might be established by pooling together the different features of the two design philosophies and principles, we came up with the MMD model. We discussed and qualified, theoretically and through illustrating comments from students, some of the specific features and potentials of the learning architecture that it implies: the power of the Now as the stage on which presence is strong, learning happens, and shared insight is collaboratively constructed; the wide resource concept, enabling learners to establish ownership and authenticity; the possibility of reflective meta-learning stimulating awareness and creation of identities through retrospection of past dialogue; the teacher-student roles bringing the learners in the centre of the stage, and the teacher to what we may call a "meta-periphery" of the learning process; the emphasis on soulful learning as an inherent dimension of the MMD architecture; the transparency of the architecture itself as a resource for meta-learning; and, finally, the global learning vision and potential of the MMD architecture for building bridges across a variety of global diversities:

Those who can understand the informal yet structured, experiential yet social, character of learning — and can translate their insight into designs in the service of learning — will be the architects of our tomorrow. (Wenger, 1998, p. 225)

The challenges ahead are numerous. The very obvious one (and next step) is to implement the MMD architecture in a variety of learning contexts within networked adult education, and to further explore its power and potential, and its potential limitations, as a pedagogic design framework in practice. However, irrespective of how many obstacles may appear along the road, we firmly believe that the MMD design model represents an empowering path to pursue, in preparing the global grounds for a type of networked learning that fruitfully unfolds in the "Now," while promoting democratic steps on a world stage of meaningful, authentic, responsible, holistic, and soulful learning.

References

Bates, A. W. (1999). *Managing technological change: Strategies for academic leaders*. San Francisco: Jossey Bass.

Bateson, G. (1976). *Steps to an ecology of mind*. Chicago: The University of Chicago Press.

Bhattacharya, M. (2002). *CDTL Brief, 5*(3). Retrieved February 15, 2005, from http://www.cdtl.nus.edu.sg/brief/v5n3/sec3.htm

Brecht, B. (1968). On the experimental theatre. In J. L. Calderwood & H. E. Toliver (Eds.), *Perspectives on drama* (pp. 270-277). New York: Oxford University Press.

Colaizzi, P. F. (1978) Learning and existence. In R. Valle & M. King (Eds.), *Existential-phenomenological alternatives for psychology* (pp. 119-135). New York: Oxford University Press.

Collins, T., Mulholland, P., & Watt, S. (2001). Using genre to support active participation in learning communities. *Proceedings of the First European Conference on Computer Supported Collaborative Learning (Euro CSCL 2001)* (pp. 156-164).

Collis, B. (1996). *Tele-learning in a digital world: The future of distance learning*. London: International Thomson Publications.

Collis, B., & Moonen, J. (2001, second printing, 2002). *Flexible learning in a digital world: Experiences and expectations.* London: Kogan Page.

The complete works of William Shakespeare. (1996). Hertfordshire, UK: Wordsworth Editions Ltd.

Daniels, H., & Bizar, M. (1998). *Methods that matter: Six structures for best practice classrooms.* York, ME: Stenhouse.

Gardner, H. (1983). *Multiple intelligences.* New York: Basic Books.

Harasim, L. (1999). A framework for online learning: The virtual-U. *IEEE Computer Society Journal "Computer", 32*(9), 44-49.

Harasim, L., Hiltz, S. R., Teles, L., & Turoff, M. (1995). *Learning networks.* Cambridge, MA: The MIT Press.

Kaye, A. R. (1989). Computer-mediated communication and distance education. In R. Mason & A. R. Kaye (Eds.), *Mindweave. Communication, computers and distance education* (pp. 3-21). Oxford: Pergamon Press.

Kaye, A. R. (1993). Learning together apart. In Kaye, A. R. (Ed.), *Collaborative learning through computer conferencing. NATO ASI Series* (Vol. 90, pp. 1-24). Heidelberg: Springer-Verlag.

Kirkeby, O. F. (1994): *Verden, ord og tanke. Sprogfilosofi og fænomenologi.* Aarhus: Handelshøjskolens forlag.

Koschmann, T. D. (1996). Paradigm shifts and instructional technology: An introduction. In T. D. Koschmann (Ed.), *CSCL — Theory and practice of an emerging paradigm* (pp. 1-23). Mahwah, NJ: Lawrence Erlbaum Associates Publishers.

Kozma, R. (1994). Will media influence learning? Reframing the debate. *Educational Technology Research and Development, 42*(2), 7-19.

Kozma, R., & Schank, P. (1998). Connecting with the twenty-first century: Technology in support of educational reform. In C. Dede (Ed.), *Technology and learning.* Washington, DC: American Society for Curriculum Development.

Lave, J., & Wenger, E. (1991). *Situated learning: Legitimate peripheral participation.* Cambridge, UK: Cambridge University Press.

Mason, R. D. (1993). The textuality of computer networking. In R. D. Mason (Ed.), *Computer conferencing. The last word* (pp. 23-36). Victoria, BC: Beach Holme Publishers.

Mason, R. (1998). *Globalising e-education. Trends and applications.* London: Routledge.

Miyake, N., & Koschmann, T. D. (2002). Realizations of CSCL conversations: Technology transfer and the CSILE project. In T. Koschmann, R. Hall, & N. Miyake (Eds.), *CSCL 2: Carrying forward the conversation* (pp. 3-10). Mahwah, NJ: Lawrence Erlbaum.

Ó Murchú, D. (2003). Mentoring, technology and the 21st century's new perspectives, challenges and possibilities for educators. *2nd Global Conference, Virtual Learning & Higher Education.* Oxford, UK.

Ó Murchú, D. (2005). Keynote. *CESI Conference.* Dublin, Ireland. Retrieved August 3, 2005, from http://www.cesi.ie

Ó Murchú, D., & Sorensen, E. K. (2003). "Mastering" communities of practice across cultures and national borders. In the Proceedings of the 10th International Conference on Open and Distance Learning, Madingly Hall, Cambirdge, UK. In Gasdell, A., & Tait, A. (Eds.), *Collected conference papers.* The Open University in teh East of England Cintra House.

Ó Murchú, D., & Sorensen, E. K. (2004). Online master communities of practice: Collaborative learning in an intercultural perspective. *European Journal of Open and Distance Learning, 2004/1.* Retrieved June 3, 2005, from http://www.eurodl.org/

Papert, S. (1993). *The children's machine: Rethinking school in the age of the computer.* New York: Basic Books.

Salmon, G. (2000). *E-moderating. The key to teaching and learning online.* London: Kogan Page.

Salomon, G. (1992). What does the design of effective CSCL require and how do we study its effects? *SIGCUE Outlook,* Special Issue on CSCL, *21*(3), 62-68.

Scardamalia, M., & Bereiter, C. (1996). Computer support for knowledge building communities. In T. D. Koschmann, R. Hall, & N. Miyake (Eds.), *CSCL theory and practice of an emerging paradigm* (pp. 249-268). Mahwah, NJ: Lawrence Erlbaum Associates Publishers.

Sorensen, E. K. (1993). Dialogues in networks. In P. B. Andersen, B. Holmqvist, & J. F. Jensen (Eds.): *The computer as a medium* (pp. 389-421). Cambridge, UK: Cambridge University Press.

Sorensen, E. K. (1997). *På vej mod et virtuelt læringsparadigme.* In J. C. Jacobsen (Ed.), *Refleksive læreprocesser* (pp. 78-109). Copenhagen: Forlaget Politisk Revy.

Sorensen, E. K. (2004). Reflection and intellectual amplification in online communities of collaborative learning. In T. S. Roberts (Ed.), *Online collaborative learning: Theory and practice* (pp. 242-261). Hershey, PA: Information Science Publishing.

Sorensen, E. K. & Ó Murchú, D. (2004). Designing online learning communities of practice: A democratic perspective. *Journal of Educational Multimedia, 29*(3).

Sorensen, E. K., & Takle, E. S. (2002). Collaborative knowledge building in Web-based learning: Assessing the quality of dialogue. *The International Journal on E-Learning* (IJEL), *1*(1), 28-32.

Sorensen, E.K., & Takle, E.S. (2003). Learning through discussion and dialogue in computer supported collaborative networks. *Society for Information Technology and Teacher Education International Conference*, (1), 2504-2510.

Sorensen, E.K., & Takle, G.S. (2004). Diagnosing quality of knowledge building dialogue in online learning communities of practice. *World Conference on Educational Multimedia, Hypermedia and Telecommunications*, (1), 2739-2745.

Tolle, E. (1999). *The power of the now*. London: Hodder & Stoughton.

Wenger, E. (1998). *Communities of practice. Learning, meaning and identity*. Cambridge. UK: Cambridge University Press.

Wenger, E., McDermott, R., & Snyder, W. (2002). *Cultivating communities of practice. A guide to managing knowledge*. Cambridge, MA: Harvard Business School Press.

Wilson, B. G. (1997). Reflections on constructivism and instructional design. In C. R. Dills & A. A. Romiszowski (Eds.), *Instructional development paradigms*. Englewood Cliffs, NJ: Educational Technology Publications. Retrieved March 16, 2005, from http://carbon.cudenver.edu/~bwilson/construct.html

Wilson, C. (1998). *The books of my life*. Charlotteville: Hampton Roads Publishing Company.

Endnotes

[1]   The administration of MIL takes place at Aalborg University, but the curriculum is developed and offered in joint collaboration between selected departments of five Danish universities (Aalborg University, Aarhus University, Copenhagen Business School, the Danish University of Education, and Roskilde University). The program is the result of ten years of research collaboration between groups of people from these departments.

[2]   The administration of MEd. And MA in Education takes place at Mary Immaculate College, University of Limerick Ireland. The curriculum is developed and offered by the College's Department of Education and awarded by the University of Limerick. The program is the result of many years of research collaboration between expert groups in MIC.

Chapter XI

Ms. Chips and Her Battle Against the Cyborgs:
Embedding ICT in Professional Praxis

J. P. Cuthell, Virtual Learning, & MirandaNet Academy, UK

Abstract

This chapter is written for practising teachers, and examines the institutional and individual factors that inhibit the implementation of information and communication technology (ICT) as a tool for teaching and learning. The affordances of ICT are identified, together with their contribution to attainment, creativity, and learning. The author argues that many of the obstacles to meaningful uses of ICT are embedded in the assumptions inherent in many institutional frameworks that are predicated on an outmoded industrial model that drives many school timetables, which process learners through the school machine. Individual change is easier to effect than institutional: the author provides some suggestions to liberate creative teachers from constraints of the system.

Preamble

In the book, *Goodbye, Mr. Chips,* by James Hilton, a shy, British teacher (Mr. Chipping, hence "Chips") devotes his life to teaching, after the death of his beautiful American wife. The film (1939) of the book features Robert Donat as Mr. Chips, who looks back on his long career and the people in it. And there you have it: the quintessential image of the devoted teacher, interacting with pupils and students, enabling them to excel and achieve their goals. The reality, of course, is often slightly different. Teachers are caught between the Scylla of an increasingly prescriptive curriculum, and the Charybdis of public accountability; schools are expected to pick up the shortfall of parenting and social responsibility, abandoned as parents rush out to work increasingly long hours to service mortgage and consumer debt.

And yet … and yet. The teacher still sees herself as someone who can shape and nurture a young life, foster a love of learning and a sense of self-worth that will help create a rounded individual: someone for whom education is more about kindling fires than filling empty vessels.

And then the door opens, large boxes are brought in, their contents fill the rooms, and the pupils spend all their time staring at the screens — or not, because the computers are not as good as the ones they have at home, are not as much fun, cannot be customized — but the ecology of the classroom has changed, and Ms. Chips has begun her relentless battle against the Cyborgs.

Teaching, the Existential Reality

The singer, Sting, once reflected on his previous career as a teacher, and commented that there was not too much difference between teaching and singing in a band: it was a question of standing in front of a crowd of screaming teenagers and trying to keep them entertained. That may well be unfair to one or more of the stakeholders in that relationship. Some teenagers have the ability to entertain themselves. Many of those I taught seemed to find endless hours of pleasure in staring through the window. From time to time, they would hear their name, realise that they were being asked a question and, offer up an attempt at an answer. They had to be there. I had to be there. We tried hard to be polite to one another and do what we all had to do.

What did we have to do? Simple. I had to teach. They had to learn. That was how the system worked. Then they would be tested to see how well they had learned. Ah yes: simple.

"Teachers shout at you. I don't shout at myself." (Boy, Year 7)

"When I'm taught people tell me the things I am learning, but when I am learning I do it myself." (Girl, Year 7)

"They talk, we listen." (Boy, Year 9)

"Learning is something you do for yourself. Being taught is something the teacher does for you." (Girl, Year 9) (Cuthell, 2002)

Something must be better than this, and ICT is often seen as the answer (Cuthell, 1998): teachers can be liberated from routine drudgery; learners set free to work autonomously and, in the words of Clynes and Kline (1960), routine checks and monitoring would be undertaken automatically, so that the human would be free to create, think, feel, and explore. Cyborg promises are not all they seem, however ....

Consider the Mise-En-Scène

The French critic and philosopher, Andre Bazin (1957), commented[1] that characters only exist within the mise-en-scène — the framing context and surrounding within which we view them. It is worth considering the mise-en-scène of schools, and the ways in which it shapes the characters who act out their existence within it. Never forget that this mise-en-scène is created from a series of acts and decisions, all of which have consequences for us, and nothing is immutable. Foucault (1977) reminds us that power exists because we accept it. When we no longer accede to it, we free ourselves. So, what is the mise-en-scène of schools? What are the framing contexts and surroundings? — What acts and decisions have created it, and, more importantly, how does this hand over power from ourselves, as education professionals, and young people, as potential autonomous learners (Cuthell, 2001)?

The Physical Environment

Schools and classrooms represent the physical manifestation of what is considered to be "education," from the phantasmagoria of Hogwarts, the boarding school for neophyte wizards in the Harry Potter stories, through the red, brick,

Victorian municipal schools found in towns and cities across the United Kingdom, to the featureless, neomodernist, concrete and plate glass structures thrown up across Europe in the second half of the twentieth century: all have their roots in what each generation perceives education to be. If we visit schools built since the start of this millennium we see that, although the surface features appear to be different, underlying structures are still the same.

Schools consist of a series of spaces: spaces along which people move, and spaces in which people are contained. These are called classrooms. Some classrooms, built for groups of up to 40, still function in a similar way today, although smaller classes sometimes mean that there is more space. The windows are often down one side of the room; there are rows of desks or tables; the teacher and board are usually at the front of the room.

Why am I describing the obvious? Because we accept it as the obvious, and there is no reason for it to be so. Why should we assume that this is how the environment for learning should be? If we accept that this is our mise-en-scène, then it means that the existential reality must be that E=(T+K): education is equal to one room containing one teacher and a group of kids. At appointed times, the room opens and its occupants transfer to other rooms, in which the process is repeated. The times are determined, not by whether the young people have learned anything, but rather by the timetable, that temporal resource manager of the curriculum.

Some primary and elementary schools manage to present the environment of a "home room," in which the décor and fixtures approximate some sense of normality, and where, in the best of schools, there is a sense of shared purpose, responsibility, and belonging.

In many secondary schools, however, classes ebb and flow with the changes of the timetables, and a stable learning environment is much more difficult to maintain. Teachers lead a peripatetic existence, moving from room to room, class to class, clutching the tools of their trade: pens, papers, books, files. What is regarded as a stable learning environment becomes reduced to the ability to impose relative order on a restless group of young people.

The Working Environment

Schools present a working environment (for both teachers and pupils) unlike most others. Possessions, materials, and personnel are all housed together in one room; there is limited personal workspace for educators or learners; where there is ICT equipment, it is set up for learner use, rather than educator use. What this means is that the systems are predicated on managing disruption, deviant behaviour, or external influences (Bronach, Greene, Riedl, Tashner, & Zimmerman, 2004[2]).

The reality for many teachers and their pupils (Cuthell, 2002), therefore, is that the working environment of school is where only some of the work — the interactions with learners and (maybe) colleagues, for teachers, and with one's peers, for the pupils-occurs. The rest takes place at home. Contrast this with other sectors of education in which staff have offices and their own computers; student computer clusters are both plentiful and available for long hours; and in which access to, and regular use of, e-mail Web communities and online services, is accepted as an integral part of teaching, administration, and learning.

In this context, it is not difficult to understand why ICT is not an integral element of either teacher praxis or the learning process for many schools (Cuthell, 1997).

Curriculum Fragmentation

A constant element of all phases of education is that the curriculum is specified by departments of education and examination boards, and then handed to the school to implement. This all too often results in timetables that segment the working day into appropriate units of work; testing is at regular intervals, with only limited feedback (usually of a summative nature) to learners.

What effect does this have on the ways in which ICT is used in the classroom? It creates the expectation that ICT should be used in specific, curriculum-contextualised ways, so that it supports (and records) what is being taught (and learned). Computer use by pupils is seen as part of the production process. The industrial paradigm of the production line still applies to many schools.

Consider the ways in which ICT is used in the wider world: in industry, in commerce, in peoples' lives. ICT is a pervasive thread that binds together everything that is done and made possible: cell phones; digital images; digital video entertainment; computer games; advertising; text messaging and instant communication; online working, trading and entertaining; all of these are part of the vocabulary and acts of most sectors of the population. Not everybody may have access to it, yet almost everybody is aware of it. In many schools, however, ICT is seen as a series of discrete activities (some of which, such as the use of cell phones for SMS or video, is proscribed) that do not recognise the centrality of the digital environment for most young people (Cuthell, 1999). We still tend to talk about the *integration* of ICT into the curriculum, rather than the *diffusion* of technology into the ways in which we learn and work.

Ways of Working

Work by learners is geared to producing outcomes to validate the curriculum: with educators working to fulfil their targets and statistical norms, learners are viewed as other-than-Kantian[3] means to an end. The young people, however, use ICT as a way to achieve outcomes that would otherwise be restricted or limited. In simple terms, if a learner can submit an assignment that is flawlessly laid out, proofed, and printed, then it does not proclaim "this is the work of a child" in the same way as something handwritten on lined paper. Instead of reaching for the red pen and identifying surface errors, the teacher has to engage with the ideas.[4]

What ways of working could there be? The use of collaborative, online working environments offers one route: the ability of teachers to undertake formative assessment of learners' work allows for a feedback mechanism vastly superior to traditional methods. When learners are enabled to create digital artefacts in the form of digital video, audio, Websites, or blogs, they can demonstrate their understanding of whatever information or concepts the curriculum demands. These are the young people who return to their homes to create and consume the digital content, from which all of their nonschool learning is taken, and there is no government-set pass rate here.

Ways of Learning

It goes without saying that only a restricted range of learning occurs within schools. The problem is that all too often, it is only the learning legitimated and assessed by educational institutions that is regarded as valid, and that learning is predicated on two assumptions: that it must involve failure (if everyone is successful, schools must be dumbing down), and that some types of learning and knowledge are privileged above others (hence regular questioning of the legitimacy of vocational courses or university subjects containing the word "studies"). As John Holt (Holt, 1964) once observed, if schools had to teach people how to speak, we'd have a lot of dumb kids.

"Teachers shout at you. I don't shout at myself." (Boy, Year 7)

This young man is commenting on the difference between learning in school and out of school. He recognises that compulsory education may not necessarily be the same as autonomous learning. And yet, every time he switches on his computer at home, he is engaged in learning: how to do things; how to solve problems; how to be creative; how to imagine and visualise. Comments from other learners reinforce this feeling that school learning is monodirectional, not interactive.

"When I'm taught people tell me the things I am learning, but when I am learning I do it myself." (Girl, Year 7)

"They talk, we listen." (Boy, Year 9)

"Learning is something you do for yourself. Being taught is something the teacher does for you." (Girl, Year 9) (Cuthell, 2002)

How can we build a pedagogy that incorporates everything we know about learning styles and learning theory? If we look at computers simply as artefacts (not even as tools), we find that a number of affordances are built into the system that not only support, but actively encourage the learning process. The concept of multiple intelligences (Gardner, 1983) is supported through ICT use.

Verbal/linguistic aspects of learning are strengthened through text input and output: the reading of information screens, error messages, instructions, documents, and Web texts. The ability of the learner to create text documents further supports this. When this is combined with the range of support tools built into the programs — spell- and style-checkers, wizards, style sheets, and templates — learners find a deeper level of individual support than is possible in most classes.

The intelligences Gardner refers to as logical/mathematical are supported in similar ways, particularly when data can be transformed into graphical representations to facilitate understanding. The ability to manipulate numbers in a spreadsheet, and observe the global changes effected by this — the "What If?" function — enables an understanding of numbers that, for many learners, would otherwise remain an imperfect concept.

Visual/spatial intelligences can be supported and developed in a number of ways through a range of programs and utilities designed to provide tools for creativity. Computer-assisted design (CAD) software is one such example, as are art packages, image manipulation, and video-editing software. All provide young learners with opportunities to learn and do things that their schools do not provide. However, beneficial though these are, the most fundamental impact that computer use can make is that of visualisation: using the transition from one screen to another (and the ability to replay it) as a means of recalling, contextualizing, and reinforcing the learning that has gone on.

Piaget (1953) refers to the ways in which young children point at objects to link them, through language, to concepts. This ostensiveness is an important aspect of bodily/kinaesthetic learning: when a learner moves the mouse cursor across a screen to reveal its contents, or when a drop-down menu is activated, this ostensive link between hand, eye, and brain enables the transition from object to concept.

One recurrent media stereotype of a computer user is that of a solitary figure with a limited range of social contacts. This may not necessarily be the case. The interpersonal intelligence identified by Gardner may be developed through a range of online forums but, within a school context, is more often used as a coin of exchange in social contexts, where young learners come together to discuss their use of computers, whether it be for homework, to fix faults, to go online, or to play games. In these contexts, young people discuss a shared, abstract, virtual environment: a level of intellectual activity rarely utilised within their curriculum. Collaborative work in a classroom often involves a number of learners sharing a computer to complete a specified task.

The computer environment, then, supports a constructivist approach to learning (Vygotsky, 1962). The way in which an ICT-rich environment moves the learner from parts to whole ideas and concepts, and in which learning is seen as a process, with feedback as integral to that process, involves the learner in the act of learning.

Reconceptualization

Something must be better than what often passes for education, and the utopian framing of ICT is often seen as the answer: teachers can be liberated from routine drudgery; learners set free to work autonomously; and, in the words of Clynes and Kline (1960), routine checks and monitoring would be undertaken automatically, so that the human would be free to create, think, feel, and explore. Are Cyborg promises an updated form of the Delphic Oracle ...?

Young people use their computers as vehicles for the combination of motor skills, language, and symbolic manipulation through practical activities. The software that they use, predicates a greater range of possibilities, as the activities for which it is used become more complex. Their relationship with their computers is dynamic. They are in a process of continual learning that they control. By contrast, their relationship with school systems, which they may use, is often restricted and static.

The graphical user interface (GUI) of the computer desktop environment, and the plasticity of software, present learners with the ability to innovate, and encourage them to experiment. The point-and-click environment that the learners utilise reinforces the power of ostensiveness, the operation of pointing, which in turn, reinforces learning through representation by imagery and perceptual organisation. The images are the translation into visual form of prior linguistic and mathematical rendering: learners use the icon as an entity in its own right: the "virtual reality" of the semiotics of the screen.

ICT and Learning

From an institutional perspective, however, learning must be correlated with outcomes. For a number of years, the Department for Education and Skills (DfES) in England and Wales has worked with the British Education Communications and Technology Agency (Becta) to assess the impact of ICT on attainment in schools. In general terms, its findings were that ICT can make a significant contribution to teaching and learning across all subjects and ages, inside and outside the curriculum; that ICT can provide opportunities to engage and motivate children and young people, and meet their individual learning needs; ICT can help link school and home by providing access from home to teaching and learning materials, and to assessment and attendance data; and finally, that ICT can enable schools to share information and good practice in networked learning communities.

In specific terms the studies have shown that:

ICT can have a direct, positive relationship to pupil performance — equivalent in some subjects to half a GCSE grade; shown that (at Key Stage 2 - National Curriculum tests that are taken by all children in England at the age of 11 to determine attainment levels across a range of subjects) schools with good ICT resources have better achievement than schools with poor ICT resources — even when compared with schools of a similar type, irrespective of socioeconomic circumstance, and irrespective of quality of management. (Becta, 2004)[5]

The Interactive Whiteboard Initiative (IWB)

Since 2002, the DfES has supported schools in the installation of interactive whiteboards, as part of a series of projects designed to evaluate the most effective ICT tools for learning. The IWB initiative has been accompanied by a number of qualitative research projects that focus on the ways in which classroom teachers use the technology to greatest effect.[6]

Interactive whiteboards support a range of learning styles: those identified by Gardner (1983) are embedded within the affordances of the technology: multiple intelligences of the learners can easily be built on and extended.

Interactive whiteboards, therefore, support learning that is essentially constructivist in its process. Parts of the lesson build into wholes, and the progression from one page to another in the IWB software provides learners with visual scaffolding: they can see where they have come from, and where they are going. The ability to review previous stages of the learning experience grounds learners, who are then able to perceive learning as a process. Feedback is integral to this process.

A social constructivist interpretation of IWB affordances can focus on the shared problem-solving experiences when learners and teachers work through materials collaboratively. This is the true meaning of "interactive": it is not simply the number of learners who move from their seat and physically interact with the IWB during the course of the lesson, but rather the process of interactions between learners, and with their teachers.

The visualised processes become the scaffolding of classroom learning, where the board itself serves as the background proximal zone of learning. A communal constructivist approach (Holmes, Tangney, FitzGibbon, Savage, & Meehan, 2001) to learning, where new knowledge is constructed by the students and the teacher, becomes possible when the curriculum materials are enriched by spontaneous learning activities that draw in other resources to the lesson as a response to learners' comments, suggestions, or questions. When interactive whiteboards support learning in this way, the classroom becomes a community of practice in which learning is seen as a social phenomenon: knowledge is inseparable from practice, and the ability to contribute to a community creates the potential for learning.

In summary, then, interactive whiteboards are tools that support a range of learning styles, whether learners are concrete or abstract perceivers. They are particularly powerful in reinforcing visualization or observational learning. Attention is improved, and retention is increased. Learners are more motivated and productive, and learning becomes inclusive and participatory. The use of ostensiveness, in which pointing reinforces learning, enhances visualization, enabling learners to recall the stages of learning. The use of ludic elements–playful, fun devices-means that learning becomes fun.[7]

What Can Teachers Do?

All of the foregoing has identified both institutional problems and ICT solutions. For Ms. Chips, who wants to deploy ICT within her teaching, and has to do daily battle with unforgiving systems and logistical constraints, an obvious solution is to forge strategic alliances with like-minded colleagues, and develop collaborative working practices. One of the most significant steps can be taken if a group of colleagues can share rooms and facilities. They can then begin to construct the kind of environment that they want to work in, and which they feel will optimise learning. Once this has been achieved, it is a smaller step to develop digital resources and lesson frames, and the benefits vastly outweigh the time costs. Teachers can reclaim their creativity and create materials that meet the learning needs of their pupils, rather than simply the curriculum objectives of the system. They can become digital auteurs.

Teachers' Toolboxes

What can Ms. Chips and her colleagues do, however, if ICT resources and access are still not sufficient? If they are totally determined to use ICT for their own teaching and administration, and provide as many opportunities for pupil learning as possible, then they should consider buying their own digital toolkit. For a relatively modest sum, one can purchase a laptop, wireless keyboard and mouse, and a digital projector[8]. This means that, projecting onto a conventional whiteboard (or wall) and using a wireless mouse, Ms. Chips can move around the classroom, and present her pupils with whatever interactive materials she chooses. She has her own tools and resources for her own classes and learners.

The Teacher as Learning Consultant

In schools where there is a shared vision of the ways in which ICT is diffused and deployed to create effective environments and learning opportunities for all, colleagues can collaborate to create programmes and materials that will support all of their learners, so that the concept of ICT providing individualised learning for all can be realised. ICT management strategies will be geared to this end; purchasing decisions will be on the basis of effective use, rather than restricted budget demands, and teachers and other education professionals will match resources to learner needs.

In the less-than-perfect world in which Ms. Chips has to live and work, however, the ability to work as an independent professional is dependent on immediate access to whatever materials, programmes, and display techniques are available. If she and her colleagues have their own classrooms, then their equipment becomes a fixture in their learning workshops; if every classroom shifts with the vagaries of the timetable, then she will use her technology tools in whatever way is most effective. But they will be her choices: access will not be blocked by firewalls; data will not be slow to download from a server because of limited bandwidth and inadequate routers; and, most importantly, her choice of these applications will be based on the needs of the learners, rather than the resources of the institution.

So How Could It Work?

Interactive content can be created through a range of functions in office productivity programs from Microsoft, Apple, and others. Presentation software can be used for far more than a linear delivery of content, and free software to create interactive content, such as Hot Potatoes (from Half-Baked Software: http://web.uvic.ca/hrd/halfbaked/) can be found on the web. In the UK, a range of resources are available online: TeacherNet (http://www.teachernet.gov.uk) is a useful source of information and case studies; Curriculum Online (http://www.curriculumonline.gov.uk) provides information about, and links to, a wealth of curriculum materials, many of which are free.

Curriculum and schemes of work can be downloaded from relevant sites: this means teachers can be up to date in terms of what they have to do, and they are independent. If details are needed, they are there, on the laptop, rather than being "somewhere" in a departmental file.

The latest research evidence can be found to provide professional legitimacy in the face of pedagogical challenges. Sites such as Becta provide access to the latest research findings; email lists such as the ICT Research Network (research@lists.becta.org.uk), and online communities such as MirandaNet (http://www.mirandanet.ac.uk) enable practitioners to share best practice and develop teaching into an evidence-based profession, rather than one relying on anecdote and stereotype.

Ms. Chips, with her laptop full of digital content for her classes, has materials to support a range of learning needs, from whole-class activities through to individualised work and reinforcement activities to consolidate learning.

Indeed, Ms. Chips and her learners can transcend the boundaries of the classroom through the creation of online communities of practice for her classes: workspaces, discussion boards, collaborative spaces and resource portals can all be built on a range of platforms. One of the most robust, available free worldwide, is Oracle's Think.com (http://www.think.com/en_us/), which offers a secure, online, learning environment that is fast, flexible, intuitive, and designed on constructivist learning principles.

The administrative and management tools that Ms. Chips and her colleagues need for their personal use can be proprietary, and link with the school's information management systems, or be customised from spreadsheets and databases. Indeed, the virtue of professional, collaborative work is that the same format can be shared and used among colleagues, so that data is used as a powerful diagnostic and reporting tool, rather than simply one for recording performance.

Finally, a teacher with a laptop possesses all the resources for digital publishing in a number of forms: for print, for audio, for video, as images.

Conclusion

We have seen the problems facing many teachers as they try to implement ICT as a powerful tool for learning. In most cases, these are institutional, and not intentional: simply the result of systems that are not flexible enough to contain the diverse needs of educators and learners. Even in the best of all possible worlds, resource allocation struggles to keep pace with demand. When innovative use is thrown into the equation, even the most sympathetic ICT coordinators find their time and patience stretched to breaking point. In smaller schools, especially elementary or primary, dedicated ICT support is rarely on a full-time basis.

The solution that has been proposed, that of the teacher with her own digital toolkit: laptop, materials, projector, may seem extreme. In terms of costs and benefits, however, the price of independence is a relatively small one. For not only will Ms. Chips be in control of her own teaching, and support her pupils' learning needs, but she will also have independence from the limitations of her school system. For a current (March 2005) investment of £1,000 (•1,440), Ms. Chips will have tools that should last for 3 years. Not only that: she will also have the key components for a home cinema, should she be able to find time for leisure after work. And, of course, she will have a computer for personal use.

What is needed:

- A laptop with DVD and CD drive, wireless keyboard, and wireless mouse.
- A digital projector for use in classrooms (and home).
- Internet access for free resource materials.
- A printer (maybe).

With this toolkit, the new paradigm of teacher as digital auteur takes the concept of collaborative work and learning out of the institution, and puts it back into the hands of people, both educators and learners.

Ms. Chips can now work with the Cyborgs, rather than battle against them.

References

Bazin, A. (1957). Dix personnages en quête d'auteurs. *Cahiers du Cinema, VII*(67-78), 16-30. Paris: Cahiers du Cinema.

Becta. (2005). Retrieved August 8, 2005, from http://www.becta.org.uk/page_documents/research/wtrs_bibs_leadership.pdf

Bronack, S., Greene, M., Riedl, R., Tashner, J., & Zimmerman, S. (2004). Learning lockdown: The disconnect between preservice preparation and permissible technology practice in schools. *Proceedings of Society of Information Technology in Teacher Education (SITE) 2004* (pp. 1108-1112). Norfolk, VA: AACE.

Clynes, M., & Kline, N. (1960). Cyborgs in space. In C. Gray, H. Figueroa-Sarriera, & S. Mentor (Eds.), *The Cyborg handbook* (1995) (pp. 29-33). London: Routledge.

Cuthell, J. P. (1997). Patterns of computer ownership. *Computer Education, 86*, 13-21.

Cuthell, J. P. (1998). What teachers think about IT. *Computer Education, 88*, 16-19.

Cuthell, J. P. (1999). The house that Strauss built. D.I.Y. in cyberspace: Bejeaned student bricoleurs. *Computer Education, 91*, 19-21.

Cuthell, J. P. (2001). *Virtual learning.* Aldershot, UK: Ashgate.

Cuthell, J. P. (2002). A community of learners. *Journal of Interactive Learning Research, 13*(1/2), 169-188.

Cuthell, J. P. (2004). Can technology transform teaching and learning? The impact of interactive whiteboards. *Proceedings of Society of Information Technology in Teacher Education (SITE) 2004* (pp. 1133-1138). Norfolk, VA: AACE.

Foucault, M. (1977; 1991). *Discipline and punish: The birth of the prison.* London: Penguin.

Gardner, H. (1983). *Frames of mind. The theory of multiple intelligences.* London: Heinemann.

Glover, D., & Miller, D. (2002). Running with technology: The pedagogic impact of the large-scale introduction of interactive whiteboards in one secondary school. *Journal of Information Technology for Teacher Education. 10*(3), 257-276.

Glover, D., Miller, D., Averis, D., & Door, V. (2005). The interactive whiteboard: A literature survey. *Technology, Pedagogy, and Education, 14*(2), 106-113.

Goodbye, Mr. Chips. (1939). Metro Goldwyn Mayer.

Half-Baked Software. Retrieved August 8, 2005, from http://web.uvic.ca/hrd/halfbaked/

Hilton, J. (1935). *Goodbye, Mr. Chips.* New York: Bantam Books.

Holmes, B., Tangney, B., FitzGibbon, A., Savage, T., & Meehan, S. (2001). Communal constructivism: Students constructing learning for as well as with others. *Proceedings of SITE 2001* (pp. 3114-3119). Norfolk, VA: AACE.

Holt, J. (1964; 1990) *How children fail.* London: Penguin.

MirandaNorth IWB Studies. Retrieved August 8, 2005, from http://www.mirandanorth.org.uk

North Islington Education Action Zone. (2003). Retrieved August 8, 2005, from http://www.virtuallearning.org.uk/whiteboards/An_approach_to_an_effective_methodology.pdf

Piaget, J. (1953) *The origin of intelligence in the child.* London: Routledge and Kegan Paul.

SMART Education. Retrieved August 8, 2005, from http://smarteducation.canterbury.ac.uk

Smith, H., Higgins, S., Wall, K., Miller, J. (2005). Interactive whiteboards: Boon or bandwagon? A critical review of the literature. *Journal of Computer Assisted Learning, 21*(2), 91-101.

Think.com. Retrieved August 8, 2005, from http://www.think.com/en_us/

Vygotsky, L. S. (1962). *Thought and language.* Cambridge: MIT Press.

Endnotes

[1]   "Les personnages n'existent que dans le mise en scene."

[2]   What Bronach and the other authors were referring to in their paper "Learning lockdown" were the ways in which ICT networks and systems in schools were set up, and locked down, so that the system administrators and network technicians could control as many of the activities as possible. In this way, activities not explicitly sanctioned by administrators were deemed "illegal" or impermissible. This approach assumes an overall level of incompetence on the part of most users, and malice aforethought on the part of others. The reality, of course, is that those who want to get up to no good do so anyway, whilst the vast majority of legitimate users are hugely inconvenienced.

3   A quick and dirty summary of Kant's categorical Imperative: *Treat people as ends in themselves, other than means to an end.*

4   This has its own pitfalls: many teachers are seduced by surface polish to such a degree that they fail to read the content carefully enough. In many cases, this is because they lack the technical proficiency themselves to understand what has been done. See "Why can't teachers do IT? Cognitive dissonance" In: Cuthell (2002). *Virtual earning* Ashgate Aldershot, UK

5   Examples of the evidence underpinning the ICT in Schools Programme (including reports in the ICTiS Research & Evaluation Series) may be found on Becta's ICT research website: http://www.becta.org.uk/research/index.cfm

6   A number of studies have looked at the effects of interactive whiteboards (IWB) on teaching and learning. See: Becta (2005); Cuthell (2004; 2005); Glover and Miller (2002, 2005); MirandaNorth IWB Case Studies (2005); North Islington Education Action Zone (2003); SMART Education at the University of Canterbury (2005); Smith et al. (2005).

7   Interactive whiteboards: case studies from the MirandaNet-Promethean projects, 2002 – 2005 (http://www.virtuallearning.org.uk).

8   A laptop with wireless card and appropriate specifications for less than £500; a projector for less than £470, and a wireless keyboard and mouse for £27: all for less than £1,000. They should last 3 years, which means that the cost per term for a teacher is just over £100. That price buys independence and professional autonomy, as well as one's own system for personal use. (Source: http://www.dabs.com, accessed 23.02.05).

Chapter XII

Making Sense of Technologically Enhanced Learning in Context:
A Research Agenda

Simon B. Heilesen, Roskilde University, Denmark

Sisse Siggaard Jensen, Roskilde University, Denmark

Abstract

This chapter proposes that technologically enhanced learning should be understood and evaluated by means of a combination of analytical strategies. These will allow us to analyze it both as seen from the macroanalytical or "outside" perspective of a rich social, cultural, and technological context, and from a microanalytical or "inside out" perspective of individual sense-making in learning situations. As a framework, we will be using sense-making methodology, and a model for causal layered analysis. Our area of attention will be limited to the "remediated classroom" of constructivist, net-based university education. Problematizing some common assumptions about technologically enhanced learning, the authors define 10 questions that may serve as the basis for a research agenda meant to help us understand why the many visions and ideals of the online or remediated classroom are not more widely realized and demonstrated in educational design and practice.

Introduction

Internet veterans will remember BITNET. It was so named as an acronym for "Because It's There Network" although it later came to stand for "Because It's Time Network." Both definitions reflect typical, and anything but optimal, reasons for introducing new technology, including those technologies meant to enhance teaching and learning. Today, a generation after it was first introduced, information and communication technology (ICT) is used widely and arguably successfully at all levels in the educational system. Yet, some doubts remain as to just how efficacious *technologically enhanced learning* really is. One such doubt about the match between ideals and reality was expressed in the call for chapters for this volume, and it is central to the argument in this chapter:

But why then is it — with a general agreement on expectations of technology for enhancing learning — that these visions are not realized and demonstrated more widely in educational design and practice? (Sorensen & Ó Murchú, 2004, Call for chapters)

We do not presume to be able to answer that question in detail. Instead, we will discuss how the visions and expectations that have come into being relate to the realities of the current educational scene. Our basic assumption is that expectations and visions may have been, and probably remain, too high and too bright, and we will outline a research agenda for studying some important, but easily overlooked factors that have an impact on the successful use of technologically enhanced learning. The outline of our research agenda integrates macroanalytical factors of importance with a microanalytical approach. In order to understand such a complex phenomenon as technologically enhanced learning, we propose using a combination of analytical strategies that will allow us to analyze it when seen from the "outside" together with a view allowing us to understand it from the "inside out." Below, we shall therefore introduce a frame of reference based on *causal layered analysis* and *sense-making methodology*.

In this chapter, we intend to discuss only "the remediated classroom," with its complex oral and written dialogical processes. Thus, we are not going to consider any kind of distance education that can be understood primarily as a remediation of correspondence school training. In order to avoid untenable generalizations, we will introduce some further delimitation. Geographical and cultural differences make it difficult, and indeed problematic, to compare pedagogical practices across national borders and different cultures. Thus, we will deal with one particular culture, the Danish one that both authors are part of, introducing, however, an international perspective when it is appropriate. Furthermore, we will focus on university education, using it as an example of the issues likely to be encountered in tertiary education. Not only is it the area in which we have more than a decade of firsthand experience, it is also a particularly challenging one, because university teaching typically emphasizes reflected understanding on the basis of dialogue, rather than just drills and learning by rote.

Background

Computer conferencing in Danish university courses was introduced in 1982 at the Jutland Open University (Nipper, 1989). The conferencing technology, meant to realize "third generation distance learning," was chosen in order to facilitate a dialogical form of teaching that would be immediately acceptable in a culture dominated by the tradition of the "living word." The idea of learning being an active social process dates back to the mid-19th century, and is intimately associated with the Danish Folk High Schools that offer informal education as a means of self-development to adults from all walks of life. Thus, from the very start, Open University was associated with a liberal constructivist type of pedagogy that was practiced by a few "new universities," but certainly was not mainstream in university education at the time. The choice of a teaching form reminiscent of the Folk High School was sound for pedagogical reasons, and it may also have helped legitimize and position Open University. But the choice also involved setting ideals for net-based education and technologically enhanced learning that educators have been struggling with ever since.

Seen from the perspective of this liberal constructivist tradition in Danish pedagogy, "conventional" distance learning represented a functionalist, rather than humanist approach to education. It was regarded as a learning style with low potential as to self-development through collaboration and dialogue with fellow students and teachers. To win recognition, the third-generation distance-learning educational approach therefore had to "do better" than traditional education. In remediating the classroom, researchers and practitioners in the field thus have been apt to overrate the potentials of dialogue and cooperation. This has been a tendency not only in Denmark, but also in the international literature on the online/remediated classroom. The advantages of this educational approach have often been emphasized, being that is, the online classroom enhances cooperative learning (Harasim, 1989; Harasim et al., 1995; Kaye, 1989; Pychyl, Clarke, & Abarbanel, 1999), it encourages participation and interaction (Dede, 1996; Harasim, 1989; Harasim et al., 1995; Kaye, 1989), it promotes convenience and flexibility (Harasim, 1989; Harasim et al., 1995; Kaye, 1989; McComb, 1994), it balances power (Harasim et al., 1995; McComb, 1994), it is efficient (Althaus, 1997; McComb, 1994), it prepares for new roles and computer literacy in business and society (Dede, 1996; Jensen, 2001; Jensen, & Heilesen, 2005) and, in general, e-learning provides the basis for a better life in a world where life-long learning is necessary because of the quick pace of technological and social change (Davies, 1998; Harasim et al., 1995; Keegan, 1986; McCreary & Brochet, 1992; Silvio, 2001).

Strong points, such as those mentioned above, often have been ascribed to the "nature" of a technology that also offers a potential for delocated and distributed processes, common storage, intertextuality, and more. A study of the effects of teacher discourse in computer-mediated discussions shows, however, that teacher style is one of the most important factors in determining student participation and quality of responses (Ahern, Peck, & Laycock, 1992). The study indicates that most of the technology-accredited potentials associated with the online classroom depend on the "human factors," some of the most important of which are teacher style and motivation. These findings also apply to students. This is not a new insight. Many years of research into technology and organizations, based on sociotechnical systems theory and also research on educational practices, have shown the same tendencies. The important factors determining success, even in technology-driven work settings, tasks, or learning, are human resources rather than technology, just as it is also the case in conventional education.

Although most people have been highly impressed by the learning potentials, the liberating aspects, and the mediating powers of the new technology, still, it is evident that there is no such thing as a technology-based short cut in matters of learning, especially not if it has to be based on collaboration and dialogue. In order for us to move ahead, the time has come, critically, to question how the ideals of the online classroom or computer-supported collaborative-learning strategies relate to the realities of this practice.

Frame of Reference

In this section, taking universities as an example of tertiary education, we will introduce a somewhat unorthodox frame of reference that will be used for suggesting how to supplement the more familiar approaches to the study and evaluation of technologically enhanced learning. As already mentioned briefly, it combines sense-making methodology (Dervin, Foreman-Wernet, & Lauterbach, 2003) with a method of analysis, *causal layered analysis* (CLA) that was originally developed for creating well thought-out scenarios in future studies (Inayatullah, 2002, 2003). Sense-making constitutes our point of departure, and it is an underlying theme in every question that we pose below: Does it make sense to apply technologically enhanced learning? If so, to whom and in what situation(s) does it make sense?

Such questions have, in fact, been asked for decades in various forms, ranging, typically, from the technophobic bumper-sticker slogan of "Technology is the answer! What is the question?" to utilitarian analyses of social and pedagogical needs. Drawing on the latter, constructive category, we will structure our argument below on an amplification of three typical answers (cf. the previous literature review) to the question of why we should advocate technologically enhanced learning:

- Because we want to rationalize and modernize a vital, but inefficient and labour-intensive sector;
- Because it produces better results in terms of learning; and
- Because technologically enhanced learning has liberating qualities that will enable a better life.

Any of these answers can be combined, and of course several others could be added, including the "Because changes in society necessitate it," which we will deal with very briefly. An important point, however, is that the answers can never be truly simple, and that they must be deliberated in a context.

It is our contention that dealing with an extremely complex phenomenon requires a rich context, one that facilitates analysis on several interdependent levels. This is where causal layered analysis comes in. We will not undertake an actual CLA analysis, but merely use the model as a means of indicating the main levels of analysis that need to be considered. CLA defines four levels of analysis: "litany," "systemic," "worldview," and "myth and metaphor." The first three of these will be used for asking questions about the sense-makings of technologically enhanced learning, primarily from a macroanalytical perspective, or "from the outside." The fourth level of CLA-analysis requires that we change our analytical optic to also asking questions about sense-making processes when seen from the "inside out," and framed by the sense-making methodology.

By *litany* is meant issues, as presented in the press and in popular literature. It is not just hype, although hype has something to do with it. A common assumption in litany is an almost deterministic acceptance of *irresistible external driving forces*. Thus, in this line of thinking, the information society is going to happen to us, and we have to find ways to respond, including developing an educational system geared to the new realities. This would be exactly the rationale for sense-making in terms of the answer "Because changes in society necessitate it." Of course this argument occurs from time to time, but believing firmly that the "future(s)" always is the outcome of choice, we see no point in exploring in detail what is, or should be, a political debate. But we should note two consequences of the approach. Firstly, policies advocating changes to the educational system, on the premise of inevitable technological and social developments, are often hailed as being progressive and proactive, and that they are a notable motivating factor for the development of technologically enhanced learning, since goodwill and funding are often tied to implementations of them. Secondly, that change management based on a planned, centralized, top-down approach tends to produce (limited) short-time results, rather than long-term effects.

The *systemic* level is extremely complex, comprising social, cultural, economic, historical, physical, and technological factors. Most studies of the effects and promises of technologically enhanced learning emphasize this level of analysis that, in contrast to litany, involves *deliberate choice*. A choice that would normally be based on one of the earlier stated reasons for wishing to introduce technologically enhanced learning. Universities, faculty, administrators, students, their relatives, their employers, government, and political parties are the most important parties involved, and the relations between them can be many, diverse, and intricate.

On a detailed level, studies of technologically enhanced learning normally focus on teacher-student relations defined by syllabus, curriculum, and pedagogy, as well as the social context of the learning experience. Contributing factors can be found in, for example, student/employer economic and social relations, student/relatives social relations, and student/teacher/institution technological and physical (time and space) relations. On a more general level, systemic studies tend to emphasize, for example, sociocultural relations between employees and their companies/institutions; the physical and technological relations between providers and consumers of education; and the economic relations between government, universities, and faculty.

All these aspects are, of course, essential to the study of education involving technologically enhanced learning. The question here is whether they are also sufficient to explain the successes and the evident shortcomings of technologically enhanced learning. Understandably, technologically enhanced learning is often discussed in terms of learning outcomes, the parameters for optimizing the learning experience typically being planning, platform, and pedagogy. But would things look different from another perspective, and might we be ignoring some relevant factors? We wish to bring up, as a reasoned doubt, two points:

1.   Do studies of the effects of technologically enhanced learning tend to be too simple in the sense that they either take too few factors into account on the systemic level, or that they restrict the analysis to the systemic level, thus ignoring relevant factors on other levels of analysis, notably those of worldview, and myth and metaphor?

2.   Are there relevant factors and relations on the systemic level (possibly in combination with other levels) that should be given more attention than they are commonly accorded?

The *worldview* level of the CLA-model deals with *discourse and frame of analysis*. In the present context, important factors are, for example, the discourse of life-long learning and reskilling, the social role of and cultural tradition for education, work ethics, the social and cultural norms of society. It differs from the systemic level in emphasizing *why,* rather than *how,* individuals and organizations act within the cultural and social framework. Questions to be raised at this level are whether worldview factors are taken sufficiently into account, and if so, whether they are being interpreted correctly?

Both on the systemic level and the worldview level, there are considerable cultural, social, and physical differences between regions and nations. Thus, it becomes difficult to generalize from any particular example of practice, and great care should be taken when applying an approach that works well in one context, to a new and, in some respects, different context. In short, it is not possible to deal with technologically enhanced learning as a uniform phenomenon. Of course, technologically enhanced learning can be implemented in any society, but the issues to be dealt with will depend on the structure of society, patterns of education, cultural norms, and a host of other factors.

Finally, the *myth and metaphor* level, somewhat a misnomer, examines *the gut feeling or the emotional experience* associated with the subject. Central to the analysis are personal sense-making and individual norms. This fourth level of analysis, as we interpret it, deals with the *inner life* of the individual, whereas the worldview level deals with the norms dictated by society in general, or by particular groups. The questions to be raised at this level of analysis are how significant such individual factors are to evaluating technologically enhanced learning, and how we can externalize them in such a way as to make them operational.

To analyze and externalize such inner life or sense-making processes, *sense-making methodology,* and the related interview strategy and method, *the micro moment timeline interview,* appear to be particularly well suited. The approach is based on a theory of communication influenced by phenomenology and constructionism. The concepts of the "sense-making triangle": situations, gaps, bridges, and outcomes, and the related ontology, are of vital importance to the theory and methodology (Dervin et al., 2003).

In the "worldview" of the sense-making approach, shortcomings and imperfections are emphasized in the sense that life is seen as a process of oscillations between different, and often extreme and conflicting, states and conditions. What seems to be orderly and well understood may, at the next moment, be experienced as chaotic. The world, and our understanding and knowledge of the world, are thus, both orderly and chaotic, but never complete or perfect. Humans therefore struggle onward in their lives from one situation and context to another, while continuously facing the gaps of the imperfections of the world and their lives. Of particular interest to this methodology is the "fact" that we always try to strategize and bridge the gaps we are facing in order to be able to move on. This strategizing and bridging activity is at the heart of sense-making processes, and also at the very focus of the micromoment time-line interview. Situations rich in such bridging activities are therefore, questioned in-depth in such interviews, and mapped in detail moment by moment to identify the "hows" of sense-making strategies.

Questioning Old Answers

We have outlined three likely answers to why it makes sense to pay attention to technologically enhanced learning. All three of them are based on a preunderstanding of technology being a positive and facilitating factor in society, and in the life of the individual. Also, they point to goals that can be realized, and subsequently evaluated. Such implementation and reflection, more often than not, would involve rather few parameters. To use the CLA terminology, the familiar statements that we will be using to structure the following argument focus mainly on the systemic level. In attempting to formulate questions for a research agenda, our goal is to supplement the inquiry not only at the systemic level, but also, importantly, to expand it by demonstrating the significance of focusing more on the other levels of analysis.

Rationalizing and Modernizing an Inefficient and Labour-Intensive Sector

Approaching technologically enhanced learning from the point of view of industrialization, it might be expected that automation and rationalization would provide us with "more for the buck." This kind of thinking was clearly present in the highly influential 1993 report from the Danish ministry of education on *Technology-supported learning* (Undervisningsministeriet, 1993) that helped initiate large-scale experiments with technologically enhanced learning, particularly under the auspices of Denmark's National Information Centre for Technology-supported Learning (1995-2000, http://www.ctu.dk/). It should be added that the initial political enthusiasm has grown into a more sober understanding of e-learning as a complex phenomenon not likely to be equally successful in all areas of the educational system. Also, internationally, it is being realized that compared to conventional teaching, e-learning may, at best, be cost-effective rather than cost-efficient (OECD, 2005). Whether or not one subscribes to the idea that education should be modernized by means of technology, the notion itself, and the practical experiments with technologically enhanced learning, have had an impact on the universities and on the people who work there, as we will discuss.

Faculty Resources

Let us start by hypothesizing that remediating constructivist modes of teaching by means of ICT does not lead to rationalization at all. On the contrary, it may tax resources to an extent that is detrimental to the broader adoption of technologically enhanced learning in the academic environment.

To back this claim, we have the personal experience of many teachers, who have been involved in developing and delivering netbased courses, that it is extremely time-consuming work. Not many studies are available to document that impression, but an interesting example is offered by Collis, Winnips, and Moonen (2000) describing the inordinate amount of time spent by three instructors on a course — even considering that it was developed from scratch.

Claims of successful rationalization, by means of e-learning, are more likely to be based on tasks involving training of skills or acquisition of information. In some such cases, it may be possible to standardize instruction and reduce courses to a collection of reusable learning objects. While this approach may work well in various kinds of professional training, in certain parts of the academic community, it has given rise to fears (and litany) of commodification of education and the reduction of faculty to semiskilled workers (Noble, 1998).

If we accept that, at present, preparing and delivering net-based learning activities generally is more time-consuming than is conventional teaching, then it seems relevant to pose as a question for further research (question 1): *In what ways do resource requirements affect the teacher and his or her academic environment(s)?* The point is to understand how the actors (directly or indirectly involved) make sense of the offer, challenge, or requirement of working with technologically enhanced learning. The plural form is necessary because we need to consider two situations: One in which the teacher works in a mixed environment, teaching conventional as well as net-based/supported courses, and one in which he or she teaches only net-based courses.

On a *systemic level* of analysis, such an inquiry will have to identify the broadest possible range of measurable work conditions, that is, existing norms for the preparation and teaching of the relevant course formats, considering also the required pedagogical approach; the actual time spent on preparation; the level of activity during courses and in project supervision; the handling of intellectual property rights; the intensity of evaluation and follow-up activities such as course adjustments and dissemination of experiences. In the mixed environment of net-based and conventional teaching, it is also important to study how the teacher, as well as the academic and administrative leadership and colleagues understand and prioritise both ongoing net-based/supported activities and in-service training in the technical and pedagogical aspects of technologically enhanced learning.

A *worldview* perspective must be added to explain both what kind of work effort and work ethics are expected of university faculty, how teaching activities are viewed by society in general and by the social groups that are the clientele for traditional and open university education.

On the *myth and metaphor level* of "looking inside out," it will be important to gain insight into the self-understanding, and the motivation of the individuals who volunteer to, or are made to offer net-based and net-supported courses and supervision (see also the section "sense-making strategies"). It is also important to gain similar insights into the response of faculty (colleagues and academic officers) not (yet) actively involved in technologically enhanced learning, but indirectly affected by changing conditions at the work place, and by the hopes and fears arising from shifts in the allocation of resources, recruitment of various kinds of ICT specialists to complement the academic staff, new requirements and expectations, and so forth.

Institutional Approval

Pursuing the matter of resources on a different level and from a different perspective, we propose, as another question for further research (question 2): *To what extent do institutional attitudes determine the success or failure of technologically enhanced learning?* Our basic assumption is, simply, that if technologically enhanced learning is viewed favourably and is actively encouraged, it has a better chance of succeeding than if it is regarded by the majority of the members of a university community as a marginal activity, or if only lip service is paid to its potential.

In this area of research, we have to distinguish between litany, the actual capacity and use of resources, and academic worldviews. Policies concerning technologically enhanced learning are often forced on the institutions by the political system or by competitive needs. To take just one example of the former, it would be difficult for a university leadership openly to disagree when the Danish government reports to the OECD:

Computers are a naturally integrated part of the teaching as a tool the students can learn to operate and work with for their assignments and apply to create networks and strengthen the study environment, but also as a natural tool for the teachers as part of their teaching, knowledge sharing and creation of networks across educational institutions. (Ministry of Science, Technology and Innovation, 2004, p. 85)

The quote, apparently uncritical in accepting the blessings of ICT, is an example of the unquestioning top-down drive for the introduction of new technology in education that universities also have to deal with. Competitive needs may require the universities to present themselves to the authorities, to prospective students, and to the general public in much the same way. But a discourse favourable to the possibilities in technologically enhanced learning may also echo some of the worldviews prevalent in the society that the institutions are meant to serve.

Thus, the *systemic* level of analysis becomes important in uncovering the background necessary for answering the question. We should note that both top-down and bottom-up factors are at play. As to the first, it is relevant to study how the institution handles the trade-off between, on the one hand, investing in hardware, software, and training so as to modernize academic programs, teaching methods, and administration, and on the other hand, investing in brainware so as to guarantee high quality in teaching and research. The actual implementation of policy is, of course, important, since ICT is a prerequisite for technologically enhanced learning, but availability of ICT facilities in itself is not likely to be a safe indicator of attitudes and expectations.

As to the bottom-up approach, it is relevant to study how the university responds to initiatives involving technologically enhanced learning. How are such initiatives funded? How central are they to the university's core activities? To what extent is technologically enhanced learning represented in the total range of teaching and learning activities, including factors such as number, level, and types of courses (and other teaching activities)? What types of faculty are assigned to the activities, what kinds of training, work norms, and incentives are provided for the teachers? What incentives are offered to proponents of technologically enhanced learning? What kind of general ICT training is offered to the students? How are technologically enhanced learning activities credited?

Adoption at Universities

This issue is closely related to the previous one, but it deals with "horizontal" relations between faculty and students, rather than the hierarchical relations of leadership. It is based on the generally recognized pattern of diffusion of innovations (Rogers, 2003), plus the assumption that advanced uses of a "remediated classroom" are still not as generally accepted as are automation of administrative routines, e-mail, information search on the internet, or instructional programs such as drills and simulations.

Recognizing that it is human nature to favour the most those things that are useful, and observing technologically enhanced learning in the context of general ICT-literacy and usage, as well as taking into account that adoption of technologically enhanced learning is not uniform across the various academic fields, we may ask (question 3): *Why, for what purposes, and to what extent do (what level and professions of) university faculty and students embrace technologically enhanced learning?*

The question should be explored both from the systemic point of view of social relations, and from the "inside-out" perspective of individual sense making. As to the first, some evident subquestions are: What proportion of the teachers favour the use of technologically enhanced learning? What is their status (seniority, position, and prestige) and their teaching obligations (regular programs and/or open university, undergraduate/graduate/post-graduate teaching)? What are the mechanisms of diffusion of competencies in technologically enhanced learning among faculty and students? Does knowledge sharing about technologically enhanced learning differ from other kinds of professional knowledge sharing within the academic community? To what extent are technologically enhanced learning systems available to faculty and students on/off-campus, not just in terms of quantity, but also in terms of quality? That is: Are technologically enhanced learning system specifications and functionality acceptable, taking into consideration the economic, technological, and social conditions of the users?

As to the "inside-out" perspective, we may ask: How do faculty adopters of technologically enhanced learning fit in the academic environment as a whole? That is: How are their special competencies and ideas about teaching regarded by their colleagues in general? What are the individual reasons for taking up technologically enhanced learning? What are the attitudes among students to the voluntary and/or compulsory use of technologically enhanced learning in various contexts on/off-campus? What, if any, advantages do the students find in technologically enhanced learning? How do the students perceive the attitudes, expectations and competencies of their teachers as regards the use of technologically enhanced learning systems in courses, assignment work and project work?

Producing Better Results in Terms of Learning

In order to deal with this issue at all, there has to be agreement as to how the outcome of learning processes should be evaluated. Grades, retention, and performance are some of the measurable factors commonly used, and sometimes these are supplemented by evaluations where students more or less freely voice their sentiments. Also, when analysing academic performance, the nature of the technologically enhanced learning system, the learning goals, and the pedagogical considerations underlying the application must be taken into account.

Technologically enhanced learning is an all-inclusive term. The e-learning continuum spans from an instructivist approach, to the individual acquisition of skills and information, and all the way to a constructivist approach aiming at reflected understanding of complex matters obtained through collaborative work (Cheesman & Heilesen, 2000). At any point of this continuum, it is possible to demonstrate that e-learning seems to have an effect. Thus, it is a long established fact that drills and simulations are quite effective tools for training, and the CSCL literature is rich in examples of the benefits of collaborative work. The benefits of technologically enhanced learning are often taken for granted. That is the case in the Danish government report quoted previously, and also, importantly, in the discourse of the European Commission, which is one of the major providers of funding for e-learning projects (Reding, 2003). However, all positive claims notwithstanding, hard evidence is in short supply. As a matter-of-fact, some studies even provide negative evidence, suggesting that e-learners do not perform better than conventional learners (e.g., Collis, Winnips, & Moonen, 2000), and two major OECD studies suggest that, stripped of all the wishful thinking, e-learning may not have much of a positive effect at all:

Many practitioners using ICT in education are convinced of its benefits and could not imagine going back to a learning environment without it. There is, however, no clear evidence that ICT investments made by the public sector have resulted in improved performance of teachers and/or learners, nor that it has improved the quality and access to educational resources on the scales predicted. (OECD, 2001, p. 24)

The overwhelming view of respondents of the OECD/CERI survey was that e-learning had a broadly positive pedagogic impact. However, few were able to offer detailed internal research evidence to this effect. (OECD, 2005, p. 13)

Clearly, it is still necessary to collect more substantial evidence by asking (question 4): *In what ways and to what extent do various implementations of technologically enhanced learning in themselves provide better academic results?* By "in themselves" we wish to indicate that possible bias is likely to be found in the fact that "the remediated classroom," at present, still requires the investment of extra resources in terms of teachers, work hours, and materials; that teachers and students engaging in advanced uses of technologically enhanced learning may not be typical representatives of their social groups; and that external pressures may be brought to bear in terms of attention and expectations.

Liberating Qualities that will Enable a Better Life

As to liberating qualities, proponents of distance education, irrespective of social and pedagogical persuasion, have been arguing for decades that distance education furthers independence and autonomy, fulfils social needs, and offers the student individual freedom (see for example Desmond Keegan's 1986 survey of distance education traditions). It is worth noting that this goes for any kind of distance education, conventional and net based. Thus, it is not just technology that is decisive, but also the worldview understanding of the discourse on reskilling and life-long learning, and the "inside out" individual experience of the liberating qualities of education. Exactly what constitutes liberating qualities may vary from country to country, as does the degree to which technology is a prerequisite for realizing distance education. We will deal with the liberating qualities of technologically enhanced learning first in terms of degrees of freedom in the work situation, and secondly as they are experienced by the individual.

Flexibility

Technologically enhanced learning, it is claimed, makes it possible for students to pursue their studies independently of time and space. Thus, flexible learning is liberating for students, and just-in-time-learning at the workplace provides opportunities for reskilling and advancement. A recent Danish large-scale survey affirms that the blessings of flexibility are solidly entrenched in the worldview of adult learners (Lorentsen & Christensen, 2004). However, independence of time and place is not at all a simple matter, as we have argued in an earlier publication. Time and space relations merely change when collaborative work becomes net based (Jensen & Heilesen, 2005). In this context, we shall not argue the philosophical aspects, but rather the practical side: Do the present-day implementations of technologically enhanced learning systems actually support flexibility and independence of time? We doubt it, even though fine work on groupware design certainly is available (e.g., Strijbos, Martens, & Jochems, 2004). This leads us to ask (question 5): *How can educational technology be designed in such a way that it truly supports both the claims made about it and the way students actually prefer to work?* It is a question involving not only systemic analyses of the design and uses in a social context of software, but also the time spent mastering systems, and student performance and interaction. It is also an insight into the strategies that students develop for dealing with the various written, audio, and video formats, and, as we will discuss in the next section, more solid evidence of how teachers and students experience the net-based learning, and how it corresponds to the claims and ideals impressed upon them by social norms and discourse.

This important issue will be one of the themes in a forthcoming book, *Designing for Networked Communications: Strategies and Development* (Heilesen & Jensen, 2006). At this stage, we cannot provide the answers, but we can point out some of the evident shortcomings of the typical asynchronous e-learning systems.

We are not alone in raising doubts about hitherto accepted forms of computer-mediated communication. It has been known for years that for net-based, collaborative, project work to succeed, many conditions must be met, and that some face-to-face interaction is helpful for sustained, well-functioning, computer-supported collaborative work (Olson & Olson, 2000). This is supported by the actual situation in Danish Open University education where at the time of writing, only 13 out of 104 net-based/supported courses are being offered as pure distance education (http://www.unev.dk/, February 2005). The rest are hybrids of classroom and distance education, representing a kind of "blended learning" compatible with the Danish tradition for "the living word," but also representative of trends on the European scene where blended learning is seen as empowerment of students and teachers, and a reaction to commodified e-learning (Reding, 2003).

The written and asynchronous form of communication typical of many popular platforms is being questioned (Farmer, 2004), and there is a growing literature on the use of alternative means of communication by means of *social software,* such as IP-telephony, videoconferencing, real-time audio/video/text conferencing systems, chat, blogging, tagging, and wikis. However, we should be wary of falling into the trap of trying to solve all sorts of problems with a new technological fix, and we should not ignore the fact that synchronous systems, again, make us dependent on time if not of place, that they give rise to new kinds of cognitive overhead, and that they too may be time-consuming and confusing. Thus, speed and immediacy come at the cost of reflection and automatic documentation of learning activities.

Sense-Making Strategies

Studies meant to untangle all the interwoven threads of technologically enhanced learning would appear to be desirable if we are to develop technologically enhanced learning strategies that make sense to the participants. When developing such sense-making strategies, a change of analytical perspective is necessary if we are to understand the situation from "the mind's eye of the user" (Dervin et al., 2003). In changing the analytical perspective while introducing a microanalytical approach, we will primarily discuss the situation of the teacher, noting briefly that students experience quite similar situations.

Teachers multitask when practicing. Roles differ from one situation to the next. Being an experienced supervisor/teacher means having a range of roles in the personal and professional repertoire, and being able to alternate between them in different patterns, depending on the situation and participants, as well as subject area and content. Observing some of these situations from the "outside," they all look very much alike. But practicing *in* the situation, a teacher has to be able to *read the situation,* and thus to understand what it looks like when seen from the "inside." In all the different roles, we draw on our familiarity with learning situations, and on competencies in reading a situation. We adjust to this reading immediately and often intuitively, moderating the way in which to communicate. These competencies are not only required in a person-to-person situation, but also when reading an audience or a group of students, Teachers need to draw on this sensitivity to the specific configurations of a given situation. We need to know what happens to the "soft" qualities and to the diversity of these competencies when teachers remediate *from* physical face-to-face situations like supervising, leading group sessions, or giving lectures, *moving on to* a face-to-interface situation in front of a computer, on the Web, in CSCW/L environments (Computer Supported Collaborative Work/Learning), in chat rooms, or on e-mail lists.

In the remediated learning situation, the teacher is often in uncharted territory, frequently being unable to read the signs of the learning culture. Everything has to be stated explicitly and decided upon: nothing can be taken for granted. This is a highly time consuming situation. Building communities and developing online learning cultures become almost as important as the teaching activities. Of course to some extent, this is always the case in learning situations, the difference being, however, that using technologically enhanced learning, there is a lack of well-known repertoires, as the learning culture is new.

When supervising online, a teacher is very likely to find him/herself in breakdown situations where many automated, internalised, and intuitive personal and professional competencies to perform and practice are no longer adequate. Suddenly, there is a mismatch or a *gap* between the competence ready at hand

and the demands of the situation. Moving from the well-known situations of conventional education to unknown situations in technologically enhanced learning, *bridging* the gap very often happens by understanding the new situation in the light of the old one. Breakdown situations may occur, and frustrations very often arise when established learning cultures are transferred and remediated without strategizing this 'gapiness' of the situations. When practising as an online teacher, participants are dislocated in time and space. But time and space are not just a hollow shell "surrounding" our situated learning activities or learning cultures. They *ground* the learning culture. If a given configuration of time and space changes, so does the learning culture, so grounded. The overall question (question 6) therefore is: *In which situations, when facing "gaps" related to practice and performance, do teachers choose to bridge the experienced gaps, so as to make sense of the situation by grounding the learning culture on technologically enhanced learning?*

To sum up the different situations and conditions outlined, we will emphasize the following aspects of the remediation processes. In dislocated and distributed learning situations, the communication is open to endless loops, and there is no shared background as communication is *from* different situations, which means that there are no well-known patters of interaction and cooperation. In such breakdown situations, we need to know about the bridging strategies that teachers apply to their daily practice in order to make sense of the communication and learning activities. Subsequently, we will propose the following research question (7): *If well-known patterns of human interaction and cooperation break down in technologically enhanced learning, which are then the sense-making strategies by which teachers bridge those gaps: strategies allowing them to "read the situation" and to adjust to this reading immediately and thus to moderate the way in which to communicate?*

Building communities also means to establish cultural codes, norms, and ethics. In conventional education and learning, such relations have long traditions. Participants — teachers as well as students — know how to interpret beginning and endings of situations, when and how to speak and listen, how to interpret and deal with interruptions, tone of voices, and so forth. It is well-known that such conventions do not lend themselves to automatic remediation, and it is a subject of key importance when it comes to technologically enhanced learning (question 8): *Which are the analytical strategies by means of which we can identify different learning situations with specific requirements as to cultural code, norms, and ethics in technologically enhanced learning?* And, subsequently, focusing on sense making (question 9): *Which are the strategies developed by teachers to bridge the gaps to make sense of codes, norms, and ethics when applied in a technologically enhanced learning context?*

As implied, sense-making processes include both explicit knowledge and tacit knowing, based on the sensitivity of presence and body. In technologically enhanced learning, emphasis is mostly on explicit knowledge, because interaction is intertextual, and what is more, writing is in the "new alphabet" of the Turing-galaxy. Thus, we also need to know about (question 10): *What strategies are applied to make sense of learning environments when emphasis is on explicit knowledge, and tacit knowing is absent?* Assuming, of course, that tacit knowing is truly significant to learning activities.

Studies of net-based learning tend to confirm that the problems outlined above apply also from the students' perspective of sense making in dislocated and distributed learning situations, where participants communicate *from* different situations in a context with no shared background, and without well-known patterns of interaction and communication, with no common codes, norms, and ethics (fx. Kindred, 2000; Skalts & Brandt-Lassen, 2001). The questions raised about sense-making strategies are therefore also relevant for the students' situation, the difference being that we will have to take into consideration the questions of asymmetrical power relations and control, as students do not always decide for themselves in what way to meet, communicate, and create results.

Conclusion

In this chapter, we have argued that in order to understand why the many visions and ideals of the online or remediated classroom are not more widely realized and demonstrated in educational design and practice, it may be helpful to apply both a macroanalytical perspective and microanalytical strategies with an "inside out" view. We mean to supplement, but certainly not to replace factors (working alone or in combination) directly associated with technology, pedagogy, curriculum planning, communicative competencies, teacher-student relations, economy, and so forth: factors that are, of course, crucial for gaining a deeper insight and for making improvements.

To model such a rich context, we have used as a framework a combination of sense-making methodology and causal layered analysis to suggest some dynamically interrelated layers of analysis. In order to illustrate how such analyses can be made operational, we have discussed a number of issues that can be analysed either from a macro- and/or a microanalytical perspective, offering in the process a series of questions that can be used as a point of departure for a research agenda. To sum up briefly:

From a macroanalytical perspective, we have been identifying factors likely to be relevant for the understanding, adoption, and success of technologically enhanced learning in a broad social and cultural setting, combining systemic factors describing *how*, with worldview factors of norms, conventions, and prevalent discourses, and also the individual experience of emotions, motivation, and aspirations, all of them contributing to explaining *why*. Through this process we have formulated these five questions:

1. In what ways do resource requirements affect the teacher and his or her academic environment(s)?

2. To what extent do institutional attitudes determine the success or failure of technologically enhanced learning?

3. Why, for what purposes, and to what extent do (what level and professions of) university faculty and students embrace technologically enhanced learning?

4. In what ways, and to what extent, do various implementations of technologically enhanced learning in themselves provide better academic results?

5. How can educational technology be designed in such a way that it truly supports both the claims made about it and the way students actually prefer to work?

The macroanalytical perspective of observing from the *outside* should be closely integrated with a microanalytical approach with an *inside out* view from the mind's eye of the user. Because only if the remediation of learning situations and classrooms make sense to the participants in given situations, and in context, can we hope to enhance the use of new technology and media in learning. Recognizing that we need to know much more about such situations, we put forward another five questions:

6. In which situations, when facing "gaps" related to practice and performance, do teachers choose to bridge the experienced gaps so as to make sense of the situation by grounding the learning culture on technologically enhanced learning?

7. If well-known patterns of human interaction and cooperation break down in technologically enhanced learning, which are then the sense-making strategies by which teachers bridge those gaps, strategies allowing them to "read the situation," and to adjust to this reading immediately, and thus to moderate the way in which to communicate?

8. Which are the analytical strategies by means of which we can identify different learning situations with specific requirements as to cultural code, norms, and ethics in technologically enhanced learning?

9.   Which are the strategies developed by teachers to bridge the gaps to make sense of codes, norms, and ethics, when applied in a technologically enhanced learning context?

10.   What strategies are applied to make sense of learning environments when emphasis is on explicit knowledge, and tacit knowing is absent?

If technology is not always the answer, these two sets of questions will allow us to determine from the "outside" how context and tasks condition one another. And from the "inside" of the learning situations and the participants' perspective, we will be able to inquire if, and in what situations, technologically enhanced learning offers the best answers of all the possible answers to the questions raised.

Having outlined these two sets of questions, it is necessary to consider how they may be answered. It is beyond the scope of the present chapter to plan an actual empirical research project operationalizing the questions. However, we recommend that the answers to such questions should be grounded in in-depth empirical field studies in which the sense-making methodology of the micromoment time line interview may serve as a point of departure. Such empirical field studies can and should be applied to university education. The approach that we have outlined, of course, can also be applied to studies of other kinds of tertiary level education, at least as far it is organized in the manner of "the remediated classroom." It is also our contention that, given some necessary adaptation, the two sets of questions will also be applicable to studies on technologically enhanced learning in secondary education.

References

Ahern, T. C., Peck, K., & Laycock, M. (1992). The effects of teacher discourse in computer-mediated discussion. *Journal of Educational Computing Research, 8,* 291-309.

Althaus, S. L. (1997). Computer-mediated communication in the university classroom: An experiment with online discussion. *Communication Education, 46,* 158-176.

Cheesman, R., & Heilesen, S. B. (2000). *Internationally distributed and problem-oriented teaching and learning in a net-environment.* Paper presented at Borderless Education – A Seminar on Virtual University Initiatives, Copenhagen. Retrieved July 4, 2005, from http://hdl.handle.net/1800/776

Collis, B., Winnips, K., & Moonen, J. (2000). Structured support vs. learner choice via the World Wide Web (WWW): Where is the payoff? *Journal of Interactive Learning Research, 11*(2), 131-162.

Davies, D. (1998). The virtual university: a learning university. *Journal of Workplace Learning, 10*(4), 175-213.

Dede, C. (1996). The evolution of distance education: Emerging technologies and distributed learning. *The American Journal of Distance Education, 10*, 4-36.

Dervin, B., Foreman-Wernet, L., & Lauterbach, E. (2003). *Sense-making methodology reader: Selected writings of Brenda Dervin.* Cresskill, NJ: Hampton Press.

Farmer, J. (2004). Communication dynamics: Discussion boards, Weblogs, and the development of communities of inquiry in online learning environments. Paper presented at Beyond the Comfort Zone. *Proceedings of the 21st ASCILITE Conference*, Perth. Retrieved February 26, 2005, from http://www.ascilite.org.au/conferences/perth04/procs/farmer.html

Harasim, L. (1989). On-line education: A new domain. In R. Mason & A. Kaye (Eds.), *Mindweave. Communication, computers and distance education* (pp. 50-62). Oxford: Pergamon Press.

Harasim, L., Hiltz, S. R., Teles., & Turoff, M. (1995). *Learning networks: Field guide to teaching and learning online.* Cambridge, MA: MIT Press.

Heilesen, S. B., & Jensen, S. S. (2006, forthcoming). *Designing for networked communications: Strategies and development.* Hershey, PA: Idea Group Inc.

Inayatullah, S. (2002). Layered methodology: Meanings, epistemes and the politics of knowledge. *Futures, 34*(6), 479-491.

Inayatullah, S. (2003). Causal layered analysis: Unveiling and transforming the future. In J. C. Glenn (Ed.), *Futures research methodology* (CD-ROM). Washington, DC: The American Council for the United Nations University.

Jensen, S. S. (2001). *De digitale delegater - tekst og tanke i netuddannelse. En afhandling om hyperlinks i refleksiv praksis, der er face-to-interface (in Danish. The digital delegates – text and reflection in net based education. Thesis on hyperlinks in face-to-interface reflective practice).* Copenhagen: Multivers.

Jensen, S. S., & Heilesen, S. B. (2005). Time, place and identity in project work on the net. In T. S. Roberts (Ed.), *Computer-supported collaborative learning in higher education* (pp. 51-69). Hershey, PA: Idea Group Inc.

Kaye, A. (1989). Computer-mediated communication and distance education. In R. Mason & A. Kaye (Eds.), *Mindweave. Communication, computers and distance education* (pp. 3-21). Oxford: Pergamon Press.

Keegan, D. (1986). *Foundations of distance education*. London: Routledge.

Kindred, J. (2000). Thinking about the online classroom: Evaluating the "ideal" vs. the "real." *American Communication Journal, 3*(3). Retrieved February 28, 2005, from http://www.acjournal.org/holdings/vol3/Iss3/rogue4/kindred.html

Lorentsen, A., & Christensen, A. K. (2004). *Voksne som målgruppe for universiteternes efter- og videreuddannelse. En spørgeskemaundersøgelse om voksnes holdninger, forventninger og krav til efter- og videreuddannelse (in Danish. Adults as target group for continuing university education. A survey of adult attitudes, expectations and demands for continuing education)* (Forskningsrapport nr. 5). Aalborg: Institut for Læring, Aalborg Universitet.

McComb, M. (1994). Benefits of computer-mediated communication in college courses. *Communication Education, 43,* 159-170.

McCreary, E., & Brochet, M. (1992). Collaboration in international online teams. In A. Kaye (Ed.), *Collaborative learning through computer conferencing: The Najaden Papers* (pp. 69-87). Berlin; New York: Springer-Verlag.

Ministry of Science, Technology and Innovation (2004). *Danish universities in transition — Background reports to the OECD examiners panel 2003.* Copenhagen: Ministry of Science, Technology and Innovation.

Nipper, S. (1989). Third-generation distance learning and computer conferencing. In R. Mason & A. Kaye (Eds.), *Mindweave. Communication, computers and distance education* (pp. 63-73). Oxford: Pergamon Press.

Noble, D. F. (1998). Digital Diploma Mills: The Automation of Higher Education. *FirstMonday, 2*(1). Retrieved February 26, 2005, from http://firstmonday.org/issues/issue3_1/noble/

OECD (2001). *E-learning. The partnership challenge.* Paris: OECD Publications.

OECD (2005). *E-learning in tertiary education. Where do we stand?* Paris: OECD Publications.

Olson, G. M., & Olson, J. S. (2000). Distance matters. *Human-Computer Interaction, 15,* 139-178.

Pychyl, T. A., Clarke, D., & Abarbanel, T. (1999). Computer-mediated group projects: Facilitating collaborative learning with the World Wide Web. *Teaching of Psychology, 26,* 138-141.

Reding, V. (2003). *Is e-learning going mainstream? Opening of the Learntec Forum Karlsruhe, 4 February 2003*. Paper presented at the DN: SPEECH/03/48, Karlsruhe. Retrieved February 26, 2005, from http://europa.eu.int/rapid/start/cgi/guesten.ksh?p_action. gettxt=gt&doc=SPEECH/03/48|0|RAPID&lg=EN&display=

Rogers, E. M. (2003). *Diffusion of innovations* (5th ed.). New York: Free Press.

Silvio, J. (2001). *Virtual mobility and lifelong learning on the INTERNET*. Paper presented at the INET 2001, Stockholm. Retrieved June 30, 2005, from http://www.isoc.org/inet2001/CD_proceedings/U108/Silvio_ID108_INET2001.htm

Skalts, N. N., & Brandt-Lassen, T. (2001). *Chat - en sokratisk jordemoder – en afhandling om fænomenet dialog i chatmæsslge sammenhænge (in Danish. Chat – a socratic midwife- thesis on the phenomenon of dialogue in chat contexts)* (MPP Working paper, nr. 2001-7). Copenhagen: Copenhagen Business School.

Strijbos, J. W., Martens, R. L., & Jochems, W. M. G. (2004). Designing for interaction: Six steps to designing computer-supported group-based learning. *Computers & Education, 42*(4), 403-424.

Undervisningsministeriet. (1993). *Technology-supported learning (distance learning)*. Copenhagen: Danish Ministry of Education.

Chapter XIII

Brain-Based Learning

Kathleen Cercone, Housatonic Community College, USA

Abstract

Neuroscience research that explains how the brain learns is a dynamic field. Since the 1990s, there has been explosive growth in information about the neurophysiology of learning. A discussion of the neuroanatomy that is necessary to understand this research is presented first. Following the discussion of anatomy and physiology, current brain research is described, with particular focus on its implications for teaching adult students in an online environment. In addition, two instructional design theories (Gardner's multiple intelligence and Kovalik's integrated thematic instruction) that have a basis in neuroscience are examined. Recommendations founded on brain-based research, with a focus on adult education, follow, including specific activities such as crossed-lateral movement patterns and detailed online activities that can be incorporated into an online learning environment or a distance learning class (and face-to-face classroom) for adults. Comprehensive recommendations and guidelines for online learning design have been provided as suggestions for making maximum use of the brain-based principles discussed in this chapter.

Introduction

Neuroscience research findings are now scientifically confirming many learning theories first introduced during the educational reform efforts of the 1960s (Lackney, n.d.). Researchers have explored many different aspects of the brain, including anatomy, circulation, electrical activity, glucose metabolism, and neuronal growth. Even with the growth of scientific information, the human brain is, for the most part, still unknown, as the brain is extremely complex. The brain is the major controller of the body, similar to a computer's CPU (central processing unit). It is the information processor of the human body. The brain is capable of multitasking, and it "assembles, patterns, composes meaning, and sorts daily life experiences from an extraordinary number of clues" (Jensen, 2000, p. 12). The brain, in addition to being extremely complex, is a dynamic and adaptive system. The brain contains hundreds of billions of neurons and interneurons that produce an enormous number of neural nets, or groups of neurons working together, from which our daily experience is created (Lackney, n.d.).

The brain's activity is controlled by genetics, development, experience, culture, environment, and emotions, and it is constantly under stimulation to change (Gardner, 1999). Since the 1980s, significant scientific findings have emerged about how learning occurs. By the 1990s, the scientific community had started to increase dramatically with new information about the brain. Developments in technology have allowed researchers to see inside the brain, and visualize how the structures in the brain communicate. Common imaging techniques used by researchers include computerized axial tomography (CAT, or computerized X-rays), functional magnetic resonance imaging (fMRI), and positron emission tomography (PET). These tools have allowed scientists to learn more about the brain, and findings made through them are influencing the worlds of education, science, and medicine.

With advances in technology and knowledge about the brain, there has been the development of brain-compatible or brain-based learning. Brain-based learning is a new paradigm that has tremendous implications for educators and students. This chapter will define brain-based learning, and will provide an overview of the anatomy, brain chemistry, neuronal connections, and current neuroscience research that are important in understanding how learning occurs. Neuroscience research needs to be translated into brain-based learning strategies that can be used by educators, and instructional design theories need to be developed in response to the new brain-based information being discovered by scientists. These theories should attempt to translate the neuroscience research, and provide methods that help educators to develop instructional strategies. Following this discussion, recommendations will be made for the design and development of a distance learning or online course.

The Biology of Learning

According to Jensen (2000), brain-based learning is "learning in accordance with the way the brain is naturally designed to learn" (p. 6). Research about how the brain learns is being conducted across several disciplines, including psychology, neuroanatomy, genetics, biology, chemistry, sociology, and neurobiology (Jensen, 2000). Brain-based learning is biologically driven, and the conclusions developed to date have not been definitive. Research continues, and our understanding of brain-based learning will be subject to future changes. The brain-based learning approach is not a recipe for all learning, but it can be used to develop strategies that are based on the current available research.

Brain Anatomy, Chemistry, Structure, and Body Connections

To understand how the brain learns, a basic understanding of the anatomy and physiology of the brain is necessary. The largest portion of the brain is called the cerebrum. The cerebrum is the most highly evolved part of the brain, and is sometimes called the neocortex. Higher order thinking and decision making occurs here. The cerebrum is composed of two hemispheres that are connected by a neural highway, the corpus callosum. Information travels along the corpus callosum to each hemisphere so that the whole brain is involved in most activities. Each cerebrum is composed of four lobes: frontal, parietal, temporal, and occipital. Each lobe is responsible for specific activities, and each lobe depends on communication from the other lobes, as well as from the lower centers of the brain, to complete its jobs.

Every task that the brain completes requires communication and coordination among several of its parts. For example, use of the thumb requires input from the cerebellum, the midbrain, and the motor and sensory areas of the frontal and parietal lobes. The task of learning functions in a similar way, as multiple areas of the brain must communicate and work together for learning to occur.

The brain is composed of over 100 billion neurons that are interconnected by electrical circuits. Communication between neurons occurs as information is passed from one neuron to the next by an electrochemical process. Each neuron has an extension, the axon, which carries the electrochemical impulse to neighboring neurons (Figure 1). These axons carry information, on a one-way circuit, away from the cell body of the neuron. Axons connect with other neurons at synapses, which are connecting junctions. For example, every muscle is connected by axons to the brain. The brain initiates an impulse of energy that travels along the axon, which terminates at a synapse on the muscle and causes the muscle to perform the activity.

Axons modify and grow in response to any brain activity, such as learning. Learning puts demands on the brain, and the brain responds by developing new circuits to connect new information to current or past knowledge. According to Fishback (1999), "the creation of neural networks and synapses are what constitutes learning" (n.p.).

There are billions of neurons, and the number of synapses is more than 10,000 times the number of neurons (Hill, 2001). "A single neuron can have from a few thousand up to one hundred thousand synapses, and each synapse can receive information from thousands of other neurons. The resulting 100 trillion synapses make possible the complex cognition of human learners" (p. 74). Communication between neurons at a synapse is accomplished by the release of chemicals and electrical signals. At the synapse, an axon sends messages to the next organ or nerve by releasing hormones or neurotransmitters such as adrenaline and dopamine. These transmitters tell the organ or nerve what to do. For example, the axon of the sciatic nerve (thousands of axons bundled together in connective tissue) sends information from the brain to the legs. The sciatic nerve sends a

Figure 1. Nerve cell or neuron with synapse

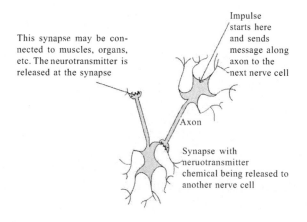

Neurons or nerve cells that connect at a synapse

neurotransmitter chemical across the synapse to the muscle that it innervates, causing a muscle contraction.

The midbrain area is deep inside the cerebrum and includes the limbic system, hypothalamus, hippocampus, and amygdala. This area works with the cerebrum, but is responsible for emotions, attention, sleep, body regulation, hormones, sexuality, and smell. This area of the brain is often called the "gatekeeper" because all incoming traffic (from the body) has to pass through it. The midbrain controls to where incoming information goes (Jensen, 2000).

The deepest part of the brain, the brain stem, is often considered the oldest and most primitive part of the brain. It is sometimes called the reptilian brain, and is responsible for our instinctual or survival behaviors. This area of the brain is the first to respond to trouble, and is the area of the "flight or fight" response. For example, the brain can "downshift" to the brain stem when a student feels threatened during a test. When that happens, the brain reacts to the situation, and it is no longer able to store or learn any information; thus, learning becomes impossible (Jensen, 2000). Both the midbrain and the brain stem will be explored further in the discussion of emotions and learning.

Neuroflexibility

Scientists once believed that the brain becomes rigid with age. It is now known that the brain is dynamic and flexible, even as one ages. In other words, the brain is plastic. The physical brain is literally shaped by experience; axonal circuits change, modify, and redevelop as human's age. "We now know that the human brain actually maintains an amazing plasticity throughout life. We can literally grow new neural connections with stimulation, even as we age. This fact means nearly any learner can increase their intelligence, without limits, using proper enrichment" (Jensen, 2000, p. 149).

In addition to adding new circuits, as axonal circuits age, pruning occurs. Pruning is the removal of connections that are no longer needed. The brain modifies its structure based on incoming information. "The brain changes physiologically as a result of experience and it happens much quicker than originally thought. The environment in which the brain operates determines to a large degree the functional ability of the brain" (Roberts, 2002, p. 282).

According to Lackney (n.d.), pruning occurs even in children, and research has shown that axons continue to grow throughout life. "You can teach old dogs a few new tricks after all. This is a huge discovery and has implications for life-long learning" (n.p.). Lackney provided an example of how this occurs in an adult who is learning how to drive a stick shift after having only driven an automatic automobile. At first, the task is frustrating and awkward for the learner, but eventually the skills become automatic. "This is a clear example of growing new neural connections and the principle of plasticity in connection with the development of body/kinesthetic intelligence" (Lackney, n.d., n.p.).

Learning is also due to input to the brain. Sensory information (e.g., aural, visual, and tactile information) enters the brain along multiple nerve receptors. Sensory input causes axons to react by budding, branching, and reaching out to other neurons, thus, leading to the development of new connections in the brain. If the information is novel, the brain needs to develop these budding new pathways. It is when an axon grows and meets up with another neuron that learning occurs. This explains why adult students need consideration of their prior experiences. Adults need to connect new information with old information. As they do this, their neural pathways change to connect new information to the older pathways already developed in the brain. Neural circuits continue to grow, even with age.

Neuronal growth, which is initiated by learning, explains scientifically what happens with assimilation and accommodation. The terms *assimilation* and *accommodation* are associated with cognitive learning theory. In assimilation, incoming information is changed to fit into existing knowledge structures (i.e., neuronal structures that already exist) (Ally, 2004). Accommodation occurs when an existing cognitive structure (i.e., current neuronal circuits) is changed to incorporate new information (Ally, 2004). Research on the neuroscience of learning is providing scientific evidence to support the learning theories that have been used for years.

Learning and the Brain

Everyone's brain is uniquely shaped by genetics, the environment, social phenomena, and experience. The interconnections, or the existing neural networks, are unique for each person. The connections between neurons are developed because of the individual's experiences, and form a "personal cognitive map" (Jensen, 2000, p. 15).

According to Leamnson (2000), our genes control what cells do and how they do it. However, after birth, chance plays a larger role than genetic code in determining whether one budding neuron will grow to another. "Genetics determines only the types of cells that get connected. The actual axonal connections are said to be "epigenetic," meaning that they are beyond, or independent of, genetic instructions" (n.p.). Epigenetic development can be seen in identical twins who have different fingerprints as well as different vein patterns, yet have identical DNA. Leamnson (2000) reported:

It happens that the connections that growing axons make upon contact with a permissive cell are often temporary. There have long been microscopic evidences of axons degenerating or withering away (just the axons, not the cells from which they grew). It is also now known that newly formed synapses are weak, or labile, and if nothing more happens the axon usually retreats, or degenerates, and the neuron starts over with a new budding axon. (n.p.)

Use is required to strengthen the neuronal connections. The more a connection is used, the larger the network grows, and the more secure the links become. The number of synaptic connections may also increase. Thus, the old adage "use it or lose it" is true of the brain (Jensen, 2000). Jensen explained that learning is defined as changing the structure of the brain. An individual's neural wiring changes as he or she learns activities. If an activity is new, the brain will respond slowly and start to develop new connections. As the activity is practiced, the pathways get more efficient, and transmission speed increases. The pathways may become permanent as the skill becomes integral to the brain, and is held in long-term memory. Neuroscience research confirms that practice improves performance.

Fishback (1999) reported that all new information is incorporated into existing neural networks. The human brain is always looking to make associations between incoming information and experience. In fact, the brain helps screen out some memory. If people remembered everything they saw, there would be too much information, and they would experience overload. "Memories are not stored intact; they are separated and then distributed in different regions of the brain" (n.p.). As information comes in and enters short-term memory, the brain must decide whether it will be consolidated and stored in long-term memory. This process of consolidation suggests that different sections of the brain must work together for learning to occur.

Brain imaging has revealed that the longer certain areas of the brain are stimulated, the better information is remembered. In addition, personal experience intensifies activation, focus, and concentration. The more elaborate a memory is (in terms of sound, touch, vision, etc.), the easier it is to access. Repetition is also important, as it causes neural connections to reactivate and increases the chance of retaining the memory (Fishback, 1999).

Neuroscience can now explain why adults need to see the link between what they are learning and how it will apply to their lives. By doing this, they strengthen their neural connections and integrate incoming information and experience into their existing wiring. Adults also need opportunities to test their learning as they go along; they should not receive only background theory and general information. They should be able to use their current neuronal circuits in order to strengthen the new pathways that are being developed as they learn new information.

The Cerebellum

Hansen and Monk (2002) researched the cerebellum and its relation to learning. The cerebellum is deep in the brain, under and behind the cerebrum. For years, it has been known that the cerebellum is responsible for coordination and balance activities, but it was only recently that the cerebellum's involvement in the learning process was discovered.

It appears that the cerebellum stores routines so that the cerebral cortex is free to deal with novel features rather than the routine....The "chunking" of routines so that the short term memory can cope with the empirically derived, short term memory space of seven bits of information is reported in all introductory texts on psychology. (Hansen & Monk, 2002, p. 346)

The "chunking" of information long described by cognitive learning theory is actually based on a physical, measurable phenomenon that researchers are just starting to understand. Many of the features of brain-based learning explain, scientifically, what is occurring at the cellular level. When learning theories evolved and developed, their principles were based on the end product, or learning. These theories were tested and researched, but they still had no measurable scientific explanation for the phenomenon being studied. It has only been recently that the underlying, physical mechanism responsible for learning has begun to be understood. As this paper continues to discuss brain-based learning, there will be scientific, measurable, objective explanations for much of what is known about learning today. However, there is still much to learn.

Preexposure and Pattern Making

Preexposure, or priming, has been shown to be important to learning. The greater the amount of a priming stimulus, the more the brain extracts and "compartmentalizes (lateralizes) the information" (Jensen, 2000, p. 81). Mechanisms of preexposure to the information presented in a course might include the following: a course description mailed out prior to the start of the course, the opportunity to talk to past participants, reading books on the subject, transparencies previewed at the beginning of the course, or a workbook (Jensen, 2000).

Preexposure to new information allows the brain to detect and create patterns of meaning. "The brain's capacity to elicit patterns of meaning is one of the key principles of brain-based learning" (Jensen, 2000, p. 82). Pattern making depends on past information or experience. The brain takes in the new information and searches for a meaningful pattern that is already in the brain with neural pathways in place. As this new information comes in, new pathways develop based on the model in the brain. "We never cognitively understand something until we can create a model or metaphor that is derived from our unique personal world" (Jensen, 2000, p. 82).

Jensen (2000) recommended the use of mind-maps, graphic organizers, advance organizers, models, or paintings of course material. The key is to get the learner to relate the information to his or her personal life. "Unless connections are made to students' prior learning, comprehension and meaning may be dramatically lessened. Before starting a new topic, ask students to discuss what they already know about the subject; do role plays or skits; make mind-maps; and brainstorm its potential value" (p. 85).

Preexposure and priming are important for adult students. Adult students need scaffolding to be provided by the instructor. *Scaffolding* is a term used by Lev Vygotsky in his social learning theory. He presented the concept of the *zone of proximal development,* which represents the gap between a student's actual level of independent problem-solving ability and the potential level that s/he could reach with knowledgeable guidance. An instructor guides students in thinking through problems and making connections. In scaffolding, the instructor first models, then coaches, then gradually withdraws help, as the learner becomes more confident and competent. Modeling by the instructor is vital, and should be part of preexposure and patterning efforts, as it allows students to see a problem solved, and then to relate new problems to a pattern that they have developed.

Biocognitive Cycles and Environmental Influences

Individuals have their own personal cycle or circadian rhythm. Research has found that everyone has an optimum time of day for learning (Jensen, 2000). The brain's right and left hemispheres alternate their cycles of efficiency. Individuals shift and use either the right or left hemisphere in a cycle during the day. This cycling activity, or lateralization, may develop a problem, and the cycle may get "stuck" in one hemisphere. This often happens in classrooms when students listen passively to a lecture. The brain's cycle may become nonresponsive, and the cycle will be altered. Jensen (2000) recommended that to "unstick" a student, an instructor should use cross-lateral activities to energize thinking and stimulate the brain to work more efficiently. An example of such an activity is using the

Table 1. Crossed-lateral activities

---

1. Take the left hand and touch the right shoulder. At the same time, touch the right hand to the left shoulder. Put both hands back with the right hand flat on the right thigh and the left hand flat on the left thigh. Alternate which hand is on top as they go to touch their shoulders again. Repeat several times. Other parts of the body can be used in this activity, rather than the shoulders. Touch with alternate hands the hips, knees, ankles, ears, and so forth.
2. Adding onto exercise number one, have them touch more than one body part, and alternate different body parts such as the shoulders and knees, and so forth.
3. Sit with feet flat on the ground. Look straight forward with hands in lap. Touch the right foot toes to the left side of the left foot. Move the right foot back to the original position. Alternate feet by putting the left toes on the right side of the right foot. Repeat several times.
4. Put hands on opposite shoulders as in exercise 1 and have students rotate to look to the right and then the left. Make sure that the whole trunk is turning and not just the neck and head.

---

right hand to touch the left leg. Another activity that could be performed in the classroom involves marching in place while patting opposite knees and touching opposite eyes, knees, elbows, and heels. Table 1 was developed to provide several examples of cross-lateral activities that could be used.

According to Jensen (2000), "Learning is best when focused, diffused, then focused again. Constant focused learning is increasingly inefficient. In fact, the whole notion of 'time on task' is in conflict biologically and educationally with the way the brain naturally learns" (p. 48).

If a student is involved actively in a learning task, it is much less likely that he or she will become "stuck," or will lateralize learning to only one hemisphere. Thus, active learning, as part of constructivist learning theory, is supported by the biology of the brain. Constructivists see learning as active rather than passive. Students construct knowledge and interpret incoming information, which is then processed by the brain to be translated into learning. The concept of lateralization also supports adult learning theories that suggest that adults need to be actively involved in the learning process.

Jensen (2000) explained that the proper environment is important for learning to occur. Color, hydration, visual stimuli, psychological stimuli, seasons, temperatures, plants, music, noise, and aromas can all influence learning. Today's classrooms are often not meeting optimal learning conditions. However, it would appear that with distance learning, the student is better able to control his or her environment. If the student is relaxed in a familiar environment, learning may be enhanced. The instructor should inform students about how to make their environments optimal for learning while using their computers at home or work.

Learning engages the entire physiology, so physical development, personal comfort, and emotional state will affect the brain. For this reason, it is important for instructors to incorporate facets of health, such as stress management, diet, and exercise, into the learning process.

Survival Mode

The brain is primarily concerned with survival, not instruction. It is the brain's prime directive to save a person's life. "The brain will concentrate on instruction that is only perceived as meaningful and only if the brain's primary survival needs have been satisfied" (Jensen, 2000, p. 14). Our brains respond to threat situations with stereotypical responses. A strong emotion, such as fear, will initiate the fight-or-flight physiological and mental response, which shuts down the higher centers of the brain (i.e., the high-level cognition centers of the cerebrum). The lower centers (i.e., the limbic system) of the brain take over to protect the individual from injury. These reflexive responses have been passed down to contemporary humans through centuries of evolution, through which the brain became the complex structure it is today. This response is reflexive; it is not controlled by conscious thought, but is in place to preserve life (Forrester & Jantzie, n.d.).

Instructors need to ensure that they do not invoke these lower centers, or learning will not occur. Stimulation is necessary for learning, but too much information can lead to overload. If a stimulus is too strong, the brain will shut down and go into survival mode. Testing can cause some students to go into "survival mode." In this situation, students will not be able to succeed, even if they have learned the material covered in the test. Thus, instructors should use multiple methods of assessment such as papers, presentations, e-portfolios, case studies, and problem-based learning tools.

Emotions

According to Jensen (2000), emotions are drivers for learning. "All learning involves our body, emotions, attitudes, and physical well being. Brain-based learning advocates that we address these multiple variables more often and more comprehensively" (p. 200). For example, those who were old enough to remember September 11, 2001, will recall exactly what they were doing on that day. This is an emotionally charged memory that most individuals will never forget. Emotion also has a strong influence on learning, and instructors should incorporate emotion into learning to make it more memorable. However, emotional stimuli must be carefully planned and balanced, in order to prevent students from shifting into "survival mode" (Fishback, 1999). If students are forced into survival mode, learning will not occur.

The amygdala is responsible for our emotions, and is concerned with survival and emotional interpretation of situations. It is responsible for bringing emotional content into memory, and plays a major role in learning. As instructors use brain-based techniques, it is important that they integrate emotions into the learning process. Reflection by the students can help this process. It is important to recognize and acknowledge the feelings and emotions that students may have. The instructor should provide personal, meaningful projects, and greater individual choice while eliminating threats, high stress, and artificial deadlines. Instructors should ensure that the resources students might need are available. It is the emotion behind the students' goals that provides the energy to accomplish them (Jensen, 2000).

In addition to establishing an emotional connection to course content, it is important to maintain a positive and supportive climate. The unconscious responses of an instructor, and the student's attitude will help determine how much learning occurs. The instructor needs to model correct behavior, and develop a nonthreatening, supportive climate. Low to moderate stress and general relaxation are best. Instructors should avoid threat-based policies and embarrassment of students (Jensen, 2000). Adult students need a climate that is collaborative, respectful, mutual, and informal in order to learn effectively. This type of environment is nonthreatening, and provides positive emotional support that helps the learning process.

Forrester and Jantzie (n.d.) reported that the limbic system plays an important part in the process of storing information as long-term memories. Activities that provide an emotionally supportive environment may have a positive effect upon the processing of information into long-term storage, and on the subsequent retrieval of those memories. The instructor should utilize group activities, cooperative learning, role-playing, and simulations, as these techniques tend to provide emotional support and an emotional context for learning. Adult students need dialogue and social interaction, as well as opportunities to collaborate with other students. These activities help provide a positive, supportive environment.

The body, emotions, brain, and mind are an integrated system (Hill, 2001). Emotions are entwined in neural connections, and emotion and cognition cannot be separated. Emotion is crucial to the storage and retrieval of information. This section has investigated the biology of learning, and has presented the scientific basis underlying current learning theory. The brain can only fully understand information if the information is meaningful. If it is meaningful, the individual will respond to events in ways that have been influenced by culture as well as personal experiences. All of these factors need to be considered in order to understand how an individual perceives and interprets incoming information. The next section will examine how learning is represented in higher order thinking skills, such as memory, metacognition, and meaning making, and will then describe the stages of brain-based learning.

Brain-Based Learning

Memory, Understanding, Thinking, and Metacognition

Memory is due to complex, multipath neuronal growth. The brain is a multimodal processor that assembles patterns, makes meaning, sorts daily life and experiences, and then processes this information. In order for information to get to the hippocampus of the midbrain, which is where long-term memory is believed to be stored, the learner needs to use the information actively to strengthen the new neural circuit. Memories are distributed throughout the cortex and are usually embedded in context. "Our brain sorts and stores information based on whether it is heavily embedded in context or in content" (Jensen, p. 222). Today's educational system often expects students to retrieve content that has been removed from context. Instructors often tell students to study from chapter 5 for a test: this is an inefficient way to teach. Using real-life simulations and contextualized situations helps students "memorize" information.

There are two different types of memory: explicit memory and implicit memory. These are further broken down into categories that are more specific. Semantic and episodic memories are considered explicit memory, or memory that was learned by effort. Implicit memory is memory that is automatically learned. It deals with nonconscious (nonmental) cognitive processing of experiences. According to Caine and Caine (n.d.), many of the insights and patterns that we grasp are the result of ongoing nonconscious processing. "Psychologists have also known for a long time that understanding is largely a consequence of deep processing. Thus, complex learning depends on a person's capacity to take charge of the processing of experience which is a matter of becoming aware of what is actually happening" (Caine & Caine, n.d., n.p.).

According to Jenkins (2000), thinking occurs when the brain accesses "prior representations for understanding" (p. 185), or when the brain tries to create a new model if one does not exist. Thinking occurs when the mind, body, and feelings are all involved. Intuition also has a role in the thinking process. Intuition is triggered by nonconscious learning that was perceived during an individual's lifetime. This is implicit memory and has no symbolic language associated with it. The basal ganglia, orbitofrontal cortex, and amygdala all contribute to intuition, since the experiences stored in these structures cannot be adequately expressed through language.

According to Hill (2001), "Consciousness refers to the ability to be self-aware and make meaning of our experiences. Consciousness can also be thought of as a sense of identity, especially the complex attitudes, beliefs, and sensitivities held by an individual" (p. 77). Sohlberg (n.d.) reported that "Nonconscious is a term that has sometimes been preferred by researchers to signify processes which are not conscious because they are by nature such that they are not available to awareness" (n.p.). In contrast, *the unconscious* is a Freudian term, by which "people usually mean mental contents that are not easily available to consciousness" (n.p.). According to Bollet and Fallon (2002), the unconscious mind is more intelligent than the conscious mind for several reasons. One reason is that the former has a greater capacity for memory. "The unconscious can hold billions and billions of bits of information, while the conscious mind can only hold 5-9 bits" (p. 41). The unconscious mind is able to perform more tasks because of this capacity. The conscious mind is more linear and sequenced (patterns), whereas the unconscious mind is emotional and nonlinear, and deals with inferences and possibilities.

It is by bringing to conscious awareness our assumptions, beliefs, habits, and practices that we begin to take charge of them and of our own learning and performance.... As we grow older we have the capacity to develop awareness and to engage in metacognitive observation. The more we can observe in our thinking, the more we can self-regulate and take charge of our own learning. (Caine, & Caine, n.d., n.p.)

Adult students should be encouraged to use their metacognitive skills to facilitate learning (Hill, 2001). Adults have many experiences and memories that act as a foundation for future learning. Adults have brains that have developed physically, with neuronal growth, in response to their experience, culture, ethnicity, personality, and political philosophy. Metacognition, or the ability to think about one's own thinking, evolves as the brain matures. Metacognition includes models of thinking, automation of conscious thought, accessing automatic processes, practice effects, and self-awareness. It also includes being aware of one's own thoughts, feelings, and actions, and their effects on others. Adult learning theories such as andragogy, transformational learning, experiential learning, and self-directed learning have elements in common, in that they encourage adults to develop metacognition.

To promote higher-level learning, metacognitive skills, or critical thinking, the online environment needs to create challenging activities that foster students' metacognitive abilities, and help them acquire meaningful knowledge. Students need the time to collaborate, interact, and reflect during the learning process (Ally, 2004). The use of self-check questions that provide feedback is a strategy that allows students to check how they are doing. By doing this self-check, students can use their own metacognitive skills to adjust the learning process if necessary.

Making Meaning or Meaningful Content

Caine and Caine (1994) asserted that the search for meaning is innate. It cannot be stopped; however, one can channel or focus it (Deveci, 2004). According to Jensen (2000), the brain is designed to seek meaning. There is a significant difference between the meaning that is gained when one memorizes material, and the meaning one achieves by developing an authentic grasp of a subject. In addition, what is meaningful to one person might be of no interest to another. According to Jensen (2000), three factors generate meaning: (a) relevance or connection with existing neural sites, (b) emotions that trigger the brain's chemistry, and (c) the context that triggers pattern making. If information is personal, emotional, and makes sense, it is meaningful. Learning and memory are context driven.

Meaning involves one's values and purpose, and it is generated when one asks questions such as "Who am I?" and "Why am I here?" The search for meaning is "survival oriented and basic to the human brain/mind. While the ways in which we make sense of our experiences change over time, the central drive to do so is life long" (Caine & Caine, n.d., n.p.). Learning should be meaningful for the student. According to Deveci (2004), "In contextual learning, students learn better when they think about what they are doing, and why they are doing it" (n.p.). The brain seeks meaningful patterns, and interprets new experiences through what is familiar. Learning should be made meaningful for students (Hill, 2001) so that they can apply and personalize new information.

Adult students need to feel that learning focuses on issues that directly concern them. They want to know what they are going to learn, how the learning will be conducted, and why it is important. Instructors should design activities for students that are interactive. The new information will assist the student in constructing new knowledge. Motivation for this process can come from a sense of personal need (Deveci, 2004). When a learner receives opportunities to develop motivation through problem-based learning or case-based learning activities, personal ownership results. Active learning is fundamental to achieving a sense of ownership. Having students draw on their previous knowledge, with the instructor acting as a facilitator in a contextual learning setting, allows the student to connect content with context, thus bringing meaning to the learning process. When connections are made with old memories and new connections are developed, new learning is fostered. "It is this initial process that constructivism has hung its hat on. In fact so much so that it would seem as though constructivism did not come about until the research was printed. On the contrary, this research simply provides a validated backdrop for which teachers can use to guide their use of this teaching approach" (Brunner, 2000, n.p.).

The Aging Brain

Neuroscience has proven that learning is possible at any age, and that cognitive growth can continue into the nineties. Normal aging and good physical health are important in the maintenance of cognitive abilities. Tests performed on individuals over 65 years of age have demonstrated that the more the brain is used and the healthier the individual, the less significantly cognitive abilities decrease over time (Anderson & Grabowski, 2003, n.p.). In order to maintain cognitive status and promote continued growth, the brain needs to be used. High levels of activity, both physical and mental, have been shown to improve cognitive changes due to age. The brain can be compared to muscle tissue that strengthens with weightlifting or activity; the brain's neuronal structures will continue to sprout and route axons as long as the brain is stimulated. Since we now know that adults can continue to learn into old age, it is important for instructional designers to be aware of the principles of brain-based learning when designing instruction for adults.

The following section of this paper describes two instructional design theories that have been developed based on neuroscience research. These theories were developed to provide instructors with methods of incorporating neuroscientific findings. Interpretation of neuroscience is necessary, and is often difficult for many who do not have training in the field. These theories have taken the findings of neuroscience research and have developed methods to assist instructors as they design learning environments.

Neuroscience and Instructional Design Theories

Instructional design theories serve as guides to professional practice, and provide different methods of facilitating learning. Instructional design theories are often built based on different learning theories. Cognitivist learning theory, although closely aligned with brain-based learning, is concerned with internal, physical changes in the learner, whereas constructivist-learning theory considers how knowledge is transferred into true learning. This author believes that the learning theory most compatible with brain-based learning is constructivism. Constructivists feel that students need to make meaning from active participation in the learning process while building personal interpretations of the world based on experiences and interactions. Constructivism also promotes the idea that learning is embedded in the context in which it is used, and that authentic tasks should be performed in meaningful, realistic settings.

Constructivist learning theory is based on the assumption that learners construct knowledge as they attempt to make sense of their experiences. What we know depends on the kinds of experiences that we have had and how we organize these into existing knowledge structures. (Boulton, 2002, p. 3)

Existing knowledge structures can be compared to the existing neural network in the brain. Learning occurs as neural connections are developed. As these neural connections develop within the existing knowledge structures in the brain, the student constructs individual meaning from information and activities. Thus, brain-based learning follows the tenets of constructivist learning theory.

There are two instructional design theories that are based on the information provided by current neuroscience research on learning. One theory is Gardner's theory of multiple approaches to learning. According to Gardner, this theory is based on human biology. The other theory is Kovalik's integrated thematic instruction (ITI) theory.

Gardner's Theory of Multiple Intelligences

According to Gardner (1999), traditional psychologists believe that there is one intelligence that is fixed, and that it may be assessed by a simple intelligence quotient (IQ) test. Gardner countered that belief by proposing that individuals have more than one intelligence. He based this assertion on neuroscience research, and developed the theory of multiple intelligences. Sherow (2000) reported:

Gardner defines intelligence as the capacity to solve problems or to fashion products that are valued in one or more cultural settings. He believes that every individual has different abilities in each area and that incorporating all intelligences in the learning process maximizes learning potential. It is important to recognize and utilize aptitudes and interests because they motivate learning; adults learn faster and better when what is to be learned is relevant and meaningful and can be applied directly to their lives and interests. (Sherow, 2000, p. 7)

Everyone possesses each of the intelligences, but to different degrees (Osciak & Milheim, 2001). For example, one person might have stronger kinesthetic learning ability than another might. Instructors need to use a variety of instructional methods and media to meet the differing needs and learning styles of students. The element of choice in learning is also important (Caine & Caine, 1990); students should have some choice in their learning activities. By having a choice, the learner can use his or her stronger intelligence to learn the necessary content. Learning styles need to be considered for all students of all ages. In any group of adults, there is a wide range of individual abilities; thus, the individualization of learning experiences is important in many situations.

Gardner's Multiple Approaches to Understanding

Gardner also developed an instructional design theory, the multiple approaches to understanding, in which he looked at "ways to foster understanding in ways that capitalize on differences in learners' intelligences" (Gardner, 1999, p. 70). Every student is bombarded by information, and each incoming stimulus gets sorted and sent to different regions of the brain in order to be interpreted in short-term memory. Once information is sent to short-term memory, "its next goal is to be employed into fitting into the category of understanding. It is through exposure and understanding that something gets committed to memory. If it does not connect through association to a previous memory or have some personal relevance to latch on to, all is lost" (Brunner, 2000, n.p.).

Gardner (1999) proposed that "Education in our time should provide the basis for enhanced understanding of our several worlds: the physical world, the biological world, the world of human beings, the world of human artifacts, the world of self" (p. 72). He saw that students need to demonstrate their understanding of important topics. Memorization, or the ability to paraphrase learning, does not constitute understanding. "Students exhibit understanding to the extent that they can invoke these sets of ideas flexibly and appropriately to carry out specific analyses, interpretations, comparisons, critiques" (Gardner, 1999, p. 81). This theory has a focus on understanding as an important type of learning outcome.

The primary goal of applying Gardner's theory of multiple intelligences should be to take advantage of the differences in students' intelligences and consider their individuality. Earle (n.d.) reported on the values upon which the theory is based, including the following:

(a)   the criticality of 'what to teach' and the variability of 'how to teach it'; (b) being able to deploy understanding (performances of understanding); (c) preparing students for valued adult roles; (d) helping students to enhance their various intelligences; (e) tailoring instruction to individual differences in students' intelligences; and (f) an approach to instruction that is not formulaic. (Earle, n.d., n.p.)

Perkins and Unger (1999) developed a similar theory about understanding. Their theory defines understanding as knowledge in thoughtful action. Their theory, like Gardner's, expands the concrete meaning of understanding. These theories place emphasis on students' appreciation of the information they are learning. "The word 'appreciate' connotes that affect is an element of deep understanding. An intersection of cognition and affect is a major thrust of the new paradigm of instructional design theories" (Gardner, 1999, p. 90). In a later article, Gardner (2003) admitted that when he formulated the multiple intelligences theory, it had an "individual-centered" bias. Gardner (2003) stated:

Most students of intelligence, however, are now coming to the realization that intelligence cannot be conceptualized, or measured with accuracy, independent of the particular contexts in which an individual happens to live, work, and play, and of the opportunities and values provided by that milieu (Gardner, 2003, n.p.)

Gardner (2003) further stated that intelligence is the result of an interaction between biology and "opportunities for learning in a particular cultural context" (n.p.). He considered the total person not only in terms of intelligence, but also as an integrated individual with motivation, personality, emotions, and will. Instructors need to respect the individual differences of every student. According to the neuroscience research discussed earlier in this chapter, everyone's brain is unique, and individuals learn in their own unique ways. Multiple intelligence theory suggests that there are observable, innate human intelligences that are based on biology. These natural intelligences are important for survival and the continuation of the species; however, the context of the individual must also be considered when discussing these intelligences.

Brain-based learning, a comprehensive approach to education, supports the notion that individual differences need to be considered in any educational setting. Although Jensen (2000) did not introduce the term *multiple intelligences*, he discussed how each person's brain is unique and individual. Everyone's intelligence is deeply entrenched in his or her context. Jensen stated that "our brain sorts and stores information based on whether it is heavily embedded in context or in content" (p. 222). Jensen described two ways that our brain deals with new information. "Information embedded in context is 'episodic' memory, which means it is stored in relationship to a particular location, circumstance, or episode; and information embedded in content is 'semantic' memory (facts), which is usually derived from reading and studying" (p. 222). Learning in context provides "more spatial and locational 'hooks' and allows learners more time to make personal connections" (p. 224). It is important for instructors to engage multiple memory pathways, and to have a focus on multiple intelligences. Ways to foster multiple intelligences include the use of real-life simulations, thematic instruction, interactive contextual learning, and a focus on multiple intelligences. When such methods are used, students activate multiple memory systems that help with learning retention.

Integrated Thematic Instructional Design Theory

The second instructional design theory that has a basis in neuroscience is the integrated thematic instruction, or ITI model, developed by Susan Kovalik. According to Dorner (n.d.), the "ITI model (Integrated Thematic Instruction) developed by Susan Kovalik carefully develops a brain-compatible learning environment, then structures curriculum to take advantage of the way the brain learns best" (n.p.).

According to Kovalik (n.d.), integrated thematic instruction is a "comprehensive model that translates the biology of learning into practical classroom and school strategies" (p. 375). Kovalik and McGeehan (1999) based their model on the "bodybrain partnership," which, as they explained, is "a coined word to reflect the collaborative activity of both brain and body for learning" (p. 376). Based on the biology of learning, they developed six learning principles: (a) intelligence is a function of experience; (b) emotions are the gatekeepers to learning and performance; (c) all cultures use multiple intelligences to solve problems and to produce products; (d) the brain's search for meaning is a search for patterns; (e) learning is the acquisition of useful mental programs; and (f) one's personality has an impact on learning. These six principles lead to nine body-brain compatible elements for instruction that serve as guides for applying the research when developing curricula and instructional strategies. Each of these principles applies to adult students as well.

The first principle involves the absence of threat, and the need to create a trustworthy environment. Instructors need to make students feel free from anxieties, and need to help students develop positive emotional associations with learning (Kovalik, n.d.). The absence of threat is necessary for the student to do reflective thinking. This principle correlates with the brain-based concept of the brain's survival response (fight or flight) to any threat situation. It is important to prevent this reflexive reaction; if it is invoked, students will "downshift" to the lower centers of the brain, and learning will not occur. Students need:

to explore the new and different and to be open to new ideas requiring confidence that one is in a safe environment, one in which mistakes and difficulty in understanding/doing something are considered just part of learning, not an opportunity for sarcasm and put-downs. (Kovalik, n.d., n.p.)

The second principle is meaningful content. Curriculum must have relevance. Learning should include examples that relate to students' lives, as such examples will assist them in understanding the information. Assignments should allow students to choose meaningful activities to help them apply and individualize information (Ally, 2004). This principle relates to the process in which axons reach out to other neurons and start to form new pathways. Prior knowledge is very important and helps to cement new information. Information that has entered the short-term memory is parked there while the student is actively using it in activities that depend on the rehearsal process (Brunner, 2000). Developing real-life context and importance to the learner is also vital. The brain looks to make associations between incoming information and experience, while also seeking patterns of meaning that depend on experiences. Two ways to develop context and importance are to consider students' prior experiences, and to encourage students to feel that they are part of a community. When an instructor uses these strategies, students' brains perceive the new information as meaningful. Students then develop a sense of well being.

The third principle is providing choices. "Provide options as to how learning will occur, considering multiple intelligences and personality preferences" (Kovalik, n.d., n.p.). Choices allow students to have control over the learning process. It is important to allow choice, as everyone's brain is unique and the correct selections are important to activate the right pathways of learning. Choices may also allow the instructor to challenge students in a supportive environment, which will decrease stress and then enhance learning.

The fourth principle is establishing a schedule that offers ample and flexible time for thorough exploration. Students need enough time to explore, understand, and use information and skills (Kovalik, n.d.). Time is needed by the brain to seek and build patterns that can be saved in long-term memory. Students need time to master a skill or concept. Time is also needed by the adult student to reflect upon and internalize information.

The fifth principle is an enriched environment that provides a learning atmosphere that reflects what is being taught. An enriched environment is a setting that provides multisensory input. The more senses that are involved in the learning experience, the longer the learned information will stay in memory. The brain is a multiprocessor of incoming sensory information that performs many operations simultaneously. A rich sensory environment enhances learning with a variety of stimuli. Examples of activities that might occur in a rich sensory environment include the following: using hands-on items, visiting sites, reading books, watching videos, and employing a variety of good references. The instructor should ensure that the room is body-compatible by avoiding clutter and distraction. Instructors should change displays and bulletin boards often, invite guest speakers, and generally increase overall sensory input. The use of multimedia in the online learning system will be discussed in greater depth later in this chapter.

The sixth principle is collaboration. Students need to work together to investigate and solve problems. "Collaboration increases understanding and improves quality of output. In the classroom, collaboration dramatically increases opportunities for the bodybrain partnership to play an active rather than passive role in learning, thus spurring physiological change in the brain" (Kovalik, n.d., n.p.). For collaboration to occur, interaction must occur. Interaction is vital to creating a sense of community, both in the classroom and online. Interaction develops a sense of community, and can promote deeper and richer learning in the adult learner. A sense of community personalizes the material for the learner and may help to contextualize information. According to Jonassen (2000), learning tasks that are meaningful, real-life tasks or activities are considered contextualized. It is this type of learning environment that puts students into learning communities.

The seventh principle is immediate feedback. It is necessary to provide coaching to promote effective teaching or learning. Immediate feedback is necessary for pattern seeking and program building. Feedback also motivates students and allows students to apply what they have learned to real-life situations (Hill, 2001).

The eighth principle is mastery/application, in which students internalize what they have learned and apply it to real-world situations. It is necessary to ensure a curriculum that allows students to acquire mental programs in order to use what they have learned in real-life situations (Kovalik, n.d.). It is important to avoid imposing perspectives on students, and to permit them to construct their own knowledge. Instructors should not make assumptions about what their students need, and should engage in communication with them. Instructors should answer questions such as "What do the students need to know and why?" and "How will this information be used?" (Imel, 2000). Online strategies should foster this transfer to real-life situations. It is possible to conduct simulations, case studies, and problem-based learning online; these approaches help students develop personal meaning, as well as contextualize the information.

The ninth and final principle is the use of movement to enhance learning. "Movement is crucial to every brain function, including memory, emotion, language, and learning" (Kovalik, n.d., n.p.). The movement principle relates to the process of lateralization, in which students' thinking can become "stuck" on one side of the brain. Cross-lateral, changing activities are important to keep the brain active on both sides, and to allow use of the entire brain.

Comparison of Gardner's and Kovalik's Theories

Gardner and Kovalik's theories serve as models for instructors. Kovalik's theory is in closer alliance with brain-based strategies than Gardner's is, since the former is based on empirical neurophysiological research. Gardner's theory also considers individual differences, diversity, and biology; however, his theory does not appear to be as heavily based on neuroscience research as Kovalik's does. Gardner based his theory on cognitive psychology and considered the brain's functions. Both of these educator-researchers have chosen to interpret the research on human learning; however, the author maintains that Gardner's multiple intelligence theory is a piece of the puzzle of Kovalik's more comprehensive theory.

The author would incorporate information from both of these theorists, especially since the author views Gardner's theory as a part of Kovalik's broader interpretation of neuroscience research. Gardner's theory has been recognized by more theorists in the literature, and provides a good foundation for understanding the individuality of students. However, it only addresses individual differences, and does not address methods of working with all the components of the "whole person" — including health, environment, context, absence of threat, respect, and pattern matching — that have been identified by Kovalik and other neuroscience researchers. It is interesting to note that both Kovalik and Gardner developed their theories in the early 1990s, when neuroscience research was in the early stages of development, and the same information was available to both authors.

It is important not only to consider individual learning differences, but also to remember that each person comes with his or her own "set of directions" and history. The positive effect of incorporating these strategies into the learning process has been proven by neuroscience research, and should be considered by instructors developing teaching strategies.

The following section of this paper will integrate ideas about specific brain-based strategies with distance learning. There are references to distance learning and online learning in the next section; the author considers the two terms to be synonymous, unless specified otherwise. These strategies are grounded in the brain-based research and instructional design theories discussed in this paper. The section will integrate and summarize the principles presented by neuroscience research, and demonstrate how they can be implemented in an adult distance-learning or online learning environment.

Brain-Based Learning and Distance Learning

Significant features to consider regarding the online learning environment for distance learning, the use of multimedia, and adult learning needs, will be presented. Ideas for developing distance-learning courses are presented that summarize and synthesize key concepts related to brain-based learning and its implications for the online environment.

Distance Learning

Distance learning and the online classroom are becoming more common today. A trend of increasing dependence on technology to assist in learning is becoming more widespread as well (Montgomery & Whiting, 2000). Understanding neuroscience research about learning and cognition can help instructors as they develop online learning environments for adults.

The brain-based learning research substantiates that learning is best facilitated when students are actively engaged in the creation of knowledge. Interaction among students (and between students and the instructor) is vital for true learning. However, technology must be used actively, and not just for transmitting information to students in a passive mode (Montgomery & Whiting, 2000). Students should not be passive recipients of information. Additionally, constructivists argue that students cannot directly learn from either teachers or technology. Students learn from thinking, and they must be actively involved in the learning process. Technologies can support learning only "if they are used as intellectual partners and tools that help learners think" (Montgomery & Whiting, 2000, p. 796).

Implications for Multimedia

Multimedia gives the instructor the ability to bring the real world to the learner with a multisensory approach, using multiple types of media simultaneously and in an integrated manner. Media might include sound, graphics, video, text, animation, or any other form of information representation. Multimedia can help motivate students and give them additional connections to their personal knowledge structures. Multimedia also helps present learning in a multimodal manner, thus allowing students to build their connections, or neural networks, in response to the material being presented. Internet and distance leaning programs can capitalize on multimedia. There is a wealth and variety of information available on the Internet, and instructors should consider this as they design their online classrooms.

Contemporary multimedia platforms allow a greater degree of learner control and more freedom for the learner to undertake self-directed exploration of the material. Such self-directed learning is likely to be more meaningful and more connected to existing knowledge structures within the learner's brain. Therefore, educators should perceive the advantages of learning programs that include multimedia presentations. (Forrester & Jantzie, n.d., n.p.)

Opportunities to self-pace and to get immediate feedback should also be built into the program or course management system. This allows students to form the correct "connections prior to reinforcing connections between new and old information incorporated within existing knowledge structures" (Forrester & Jantzie, n.d., n.p.).

Distance learning and multimedia present new challenges to today's instructors. Instructors who learned in traditional, passive classrooms need to learn new skills and ways of teaching. Today's instructors must also cope with developing new neural networks as they relate new methods of instruction to the old.

Implications for Preparing the Learner

Preexposure and scaffolding are important methods of preparing students for learning. The following are recommendations for instructors to consider as they design online learning environments: (a) use mind maps; (b) present an overview of class material before the session starts; (c) discover students' interests and backgrounds through a discussion board or students' Web pages; (d) have students set their own goals and integrate them with class goals for each unit; (e) develop a course map with hyperlinks and a site map; provide hyperlinks to related information; and (f) be a positive role model.

The instructor also needs to consider the multiple intelligences of students and provide a multimodal environment that allows learner choice. Recommendations for implementing this environment include the following: (a) use a variety of teaching tools and strategies, such as Web quests, real-life projects, role-play and design complex, and multi-sensory immersion environments; and (b) provide online students with control of navigation.

Implications for the Online Environment

Implications for the online environment include ensuring that there is positive emotional commitment by the student, with an absence of threat. Frequent feedback, mutual respect, and strong peer support are vital to maintaining a positive emotional environment. Specific recommendations for the creation of such an environment are as follows: (a) recognize that it is difficult to "unlearn" old beliefs, and allow students time to work through conflicts; (b) be aware that the learning environment may trigger past negative learning experiences for some adults; (c) use a variety of input methods, including online lectures, readings, films, videos, audio, journals, models, and pictures; (d) incorporate health education on topics such as stress management, fluid intake requirements, and movement; (e) build "attention getters" into the educational plan, and engage students in group discussions; (f) use humor and acknowledge emotions; (g) provide interactive feedback using collaborative activities and a discussion board; (h) increase rapport by developing partner learning, discussions, dialogue, and collaborative activities; (i) set up a method that allows students to express concerns in an anonymous, nonthreatening forum, such as a separate discussion board; (j) provide social opportunities for groups to interact informally; and (k) set ground rules that emphasize respect for fellow students.

It is also important to provide an authentic learning environment in which the learner can contextualize the patterns found, building on his or her previous experience. Some specific methods to achieve this include the following: (a) Get the students' attention and personalize the environment by using technology that recognizes the student's name. (b) Provide meaningful challenges by solving real-world problems. (c) Use experiments, and have students investigate using an active learning approach. (d) Present information through context, which allows the learner to identify patterns and connect them with previous experiences. (e) Use problem-based learning, collaborative/cooperative learning, project-based learning, and service learning. (f) Chunk the material; this will help students classify incoming information, and develop a pattern that they can use to build new neural networks. (g) Use students' prior knowledge. (h) Provide a model of a good performance, and then let students analyze their errors, as well as identify general patterns that underlie the concepts being studied.

Implications for the Individual

Everyone's brain is unique. It is important to promote active processing, as well as encourage reflection by the students. There should be a balance between novelty and predictability in the content. Methods of fostering active processing and reflection include the following: (a) Use a note-taking function for the students to write thoughts as they review the online material. (b) Use computer conferences such as listservs, electronic mail, bulletin boards, and MUDs (multi-user dimensions) to foster reflection. (c) Provide ways for students to engage in metacognitive reflection by using think logs, reflective journals, and group discussions within a cooperative learning setting. (d) Ensure that the student can examine the material's relevance to his or her life. (e) Provide some challenge. The brain is always looking for novelty and is responding to stimuli; thus, the environment should be stable and yet novel. (f) Use multifaceted teaching strategies in order to capitalize on the student's preferred method of learning. (g) Put collected data into a personal "scrapbook," or have each student develop an e-portfolio.

Impact of Research on Education

The principal implication of brain-based research for education is that online educators should be responsive to the latest findings on the neuroscience of learning. Scientific, objective research is proving that learning is an active and individual process. For the author, the primary question that arises from this information is this: Why aren't more educators implementing this type of training? Why are professors still "performing" in massive lecture halls? The implications of research in brain-based learning, for teaching and learning are remarkable. Educators need to be introduced to this research in a manner that allows them to understand and interpret the findings. Educators should be prudent when applying the findings of brain-based research, but at the same time, they should move forward with what is already known.

Conclusion

The underlying theme of this chapter — which has encompassed Gardner's multiple approaches to understanding, Kovalik's integrated thematic instruction, and brain-based learning — is one of diversity and individuality, which seem to be in conflict with each other. How can we teach for diversity and yet consider the individuality of each student? This is the challenge for educators, as the new brain-based research has shown the importance of diversity and individuality as characteristics of students.

Knowledge of how the brain learns provides a basis for achieving a better understanding of how adults learn. Neuroscientific research has found evidence of the brain's plasticity and ability to respond throughout life, and it has shown that adults continue to learn throughout the lifespan. Emotion and sensory experiences are integral to learning. Meaning making, morals, consciousness, and associations with others are critical to adult learning.

Two instructional design theories have been discussed that are based on the biology of learning. Kovalik's ITI theory is the most comprehensive theory, in that it incorporates much of the current research on brain-based learning. Researchers continue to discover more about how the brain learns. As new information is discovered, new learning theories will be developed. It is important to understand that research on the brain is dynamic and is rapidly providing new information. Instructional designers, as well as instructors, need to understand that the information available today may change in the near future. Instructors and instructional designers should continue to question what the theories say as new information emerges.

In the last 30 years, researchers have developed new technological tools to discover a completely unknown territory, the brain. Brain-based learning is closely aligned with the constructivist theory of learning, as well as with current information available on adult learning. Brain-based research has provided facts and objective information to support how instructors teach. The way in which the learning process is employed will have the largest impact on students' learning. A paradigm shift to constructivism that supports new instructional design and learning theories is substantiated by the research that has been presented in this paper. Learning is the beginning of discovery. Educators should consider integrating brain research into teaching strategies as learning theories continue to be developed, refined, and implemented.

Brain-based research is validating the assertion that learning is individual and unique. This implies that current practices such as standardized materials and instruction may, in fact, diminish or inhibit learning. However, instruction should not be based solely on neuroscience. Brain-based learning provides new directions for educators who want to achieve more focused, informed teaching. With additional research in brain-based approaches, there may better options for those struggling with learning. Since no two people have had the same experiences that modify neural networks, the potential for cognitive differences among individuals is huge. Brain-based research needs to be interpreted for educators so that they can utilize this information in the classroom. It is vital for the educator of tomorrow's students to understand the importance and implications of brain-based research.

References

Ally, M. (2004). Foundations of educational theory for online learning. In T. Anderson & G. Sanders (Eds.). *Theory and practice of online learning* (pp. 1-31). Athabasca, Canada: Athabasca University. Retrieved March 1, 2004, from http://cde.athabascau.ca/online_book/index.html

Anderson, S. W., & Grabowski, T. J. (2003, June). *Memory in the aging brain.* Retrieved April 13, 2004, from http://www.vh.org/adult/provider/neurology/memory

Bollet, R. M., & Fallon, S. (2002). Personalizing e-learning. *Education Media International,* 40-44. Retrieved January 11, 2004, from http://www.tandf.co.uk/journals

Boulton, J. (2002, February 26). *Web-based distance education: Pedagogy, epistemology, and instructional design.* Retrieved February 22, 2004, from http://www.usask.ca/education/coursework/802papers/boulton/boulton.pdf

Brunner, J. (2000). *Brain-based teaching and learning: A shift from content to process.* Retrieved January 15, 2004, from http://hale.pepperdine.edu/~jabrunne/brain/bbl.htm

Caine, R. N., & Caine, G. (1990, October). *Understanding a brain-based approach to learning and teaching.* Retrieved September 14, 2003, from http://poncelet.math.nthu.edu.tw/chuan/note/note/brain-based

Caine, R. N., & Caine, G. (n.d.). *Principles wheel.* Retrieved January 22, 2004, from http://cainelearning.com/pwheel/

Deveci, T. (2004). *Developing Teachers.com.* Retrieved February 2, 2004, from http://www.developingteachers.com/articles_tchtraining/brain2_tanju.htm

Dorner, K. (n.d.). *Using technology in a brain-compatible learning environment*. Retrieved January 15, 2004, from http://www.lr.k12.nj.us/ETTC/archives/dorner.html

Earle, A. (n.d.). *Operationalisation of pedagogical theory in metadata*. Retrieved January 30, 2004, from http://www-jime.open.ac.uk/2001/earle/earle-12.html

Fishback, S. J. (1999). Learning and the brain. *Adult learning, 10*(2), 18-23.

Forrester, D., & Jantzie, N. (n.d.). *Learning theories*. Retrieved January 14, 2004, from http://www.acs.ucalgary.ca/~gnjantzi/learning_theories.htm

Gardner, H. E. (1999). Multiple approaches to understanding. In C. M. Reigeluth (Eds.), *Instructional-design theories and models: A new paradigm of instructional theory* (Volume II, pp.69-89). Mahwah, NJ: Lawrence Erlbaum.

Gardner, H. (2003). *Intelligence in seven steps*. Retrieved February 20, 2004, from http://www.newhorizons.org/future/Creating_the_Future/crfut_gardner.html

Hansen, L., & Monk, M. (2002). Brain development, structuring of learning and science education: Where are we now? A review of some recent research. *International Journal of Science Education, 24*(4), 343-356.

Hill, L. H. (2001). The brain and consciousness: Sources of information for understanding adult learning. *New Directions for Adult and Continuing Education, 8,* 73-81.

Imel, S. (2000). Contextual learning in adult education. *Practice Application Briefs, 12*. Retrieved May 10, 2004, from http://cete.org/acve/pab.asp

Jensen, E. (2000). *Brain-based learning*. San Diego, CA: The Brain Store.

Jonassen, D. H., & Land, S. M. (Eds). (2000). *Theoretical foundations of learning environments*. Mahwah, NJ: Erlbaum.

Kovalik, S. (n.d.). *ITI overview*. Retrieved January 21, 2004, from http://www.kovalik.com/printer/overview_printer.htm

Kovalik, S. J., & McGeehan, J. R. (1999). Integrated thematic instruction: From brain research to application. In C. M. Reigeluth (Ed.), *Instructional-design theories and models: A new paradigm of instructional theory* (Volume II, pp. 371-396). Mahwah, NJ: Lawrence Erlbaum.

Lackney, J. A. (n.d.). *Twelve design principles based on brain-based learning research*. Retrieved January 13, 2004, from http://www.designshare.com/Research/BrainBasedLearn98.htm

Leamnson, R. (2000). Learning as biological brain change. *Change, 32*(6), 34-41.

Merriam, S. B., & Caffarella, R. S. (1999). *Learning in adulthood* (2nd ed.). San Francisco: Jossey-Bass.

Montgomery, L., & Whiting, D. (2000). *Teachers under construction: Incorporating principles of engaged and brain-based learning into constructivist "Technology in Education" program.* Paper presented at the meeting of the Society for Information Technology & Teacher Education International Conference, San Diego, CA.

Osciak, S. Y., & Milheim, W. D. (2001). Multiple intelligences and the design of Web-based instruction. *International Journal of Instructional Media, 28*(4), 355-361.

Perkins, D. N., & Unger, C. (1999). Teaching and learning for understanding. In C. M. Reigeluth (Ed.), *Instructional-design theories and models: A new paradigm of instructional theory* (pp. 91-114). Mahwah, NJ: Lawrence Erlbaum.

Roberts, J. W. (2002). Beyond learning by doing: The brain compatible approach. *The Journal of Experiential Education, 25*(2), 281-285.

Sherow, S. (2000). *Adult learning theory and practice index.* Retrieved February 2, 2004, from http://www.ed.psu.edu/paliteracycorps/hied/course_materials/Adult_Learning_Theory_and_Practice.pdf

Sohlberg, S. (n.d.). *Concepts of the unconscious.* Retrieved June 6, 2004, from http://www.psyk.uu.se/hemsidor/staffan.sohlberg/

About the Authors

Elsebeth Korsgaard Sorensen is a senior lecturer in ICT & learning in the Department of Communication, Aalborg University, Denmark. She is head of the online master's programme in ICT & learning, offered collaboratively by five Danish universities. Her international research focuses on collaborative dialogue and knowledge building online, pedagogical design and delivery of networked learning, and implementation of electronic portfolios as reflective tools in online learning processes. Sorensen presents her research at international conferences, frequently as keynote speaker, and she has published extensively in international journals and books within the field. She serves on the editorial board of several international journals, and on program committees of international conferences within the field. A list of selected publications may be found at http://www.kommunikation.aau.dk/ansatte/es/.

Daithi Ó Murchú is an all-gaelic, elementary school principal teacher. In 1996, he was awarded his MSt from Trinity College, Ireland. Following his PhD in technology and linguistics, and subsequently in elementary education and e-learning, he was elected executive vice-president of human language and technology with SITE (USA). As a cultural and technology expert with the EU's MyEurope schools, he collaborated with international universities on their e-learning and teacher education programmes. Seconded to MIC, University of Limerick, Daithí continued to work internationally in Education and Technology. He is presently collaborating with AAU, Denmark and is national director of Gaelic methodologies in e-learning environments with Hibernia College, Ireland.

* * *

Jørgen Bang is head of Department of Information and Media Studies, University of Aarhus, Denmark. Internationally, he has participated in several EU projects, and was head of the Nordic Forum for Computer Supported Learning (1991-1995). Since 2002, he has been president of the European Association for Distance Teaching Universities (EADTU). He has published books and articles within the fields of Nordic literature, media reception, net-based communication, learning theory, and learning technology.

Tony Carr is the staff development coordinator at the Centre for Educational Technology at the University of Cape Town, South Africa. As a practitioner and researcher, his interests include staff development, communities of practice, and online collaboration. His postgraduate qualifications include an MA in economics education from the Institute of Education, University of London, and a postgraduate diploma in continuing education and training from City University in the UK.

Kathleen Cercone is a licensed physical therapist who has practiced for 28 years specializing in orthopaedics and neurology. She was awarded a BS in physical therapy from Columbia University and an MS in exercise physiology from Southern Connecticut State University. She is currently attending Capella University for her PhD in education, with a concentration in instructional design for online learning. Cercone is a tenured associate professor of biology at Housatonic Community College in Connecticut, USA. She is also the coordinator of the physical therapist assistant program, as well as having taught in the program.

Erik Champion currently teaches multimedia and games design in the information environments programme at the University of Queensland, Australia. His PhD project, Cultural Immersion in Virtual Places (part of a Lonely Planet and Australian Research Council SPIRT grant) evaluated "cultural presence" in virtual heritage environments using game-style interaction. Papers from this work have appeared in the Proceedings of GRAPHITE 2002, Presence 2002, ACADIA 2002, CAAD Futures 2003, IE2004, SAHANZ 2004, VSMM 2003 & VSMM 2004, and in journals such as *ACM Computing and Society*, and *Traffic*.

Glenda Cox has been working on projects in the use of technology in teaching and online collaboration since June 2000. Initially, her focus was on the use of qualitative research software, and finding a content analysis tool to analyse synchronous and asynchronous online conversations in courses at the University of Cape Town, South Africa. She has been involved in humanities staff development workshops since the end of 2003. Her role in this project is focused on researching the take-up and effect of these workshops. She is also working in a mentor role with academics who have attended the workshops and started new technology interventions.

J. P. Cuthell is the research and implementation director for the MirandaNet Academy, UK. He has developed practice-based research accreditation for teachers based on ICT classroom projects. Past MirandaNet projects have investigated the effect of laptop computers on students and teachers at home and in school, and action research projects, with teachers evaluating the impact of interactive whiteboards on teaching and learning, now extended to involve schools and educationalists in Mexico, China, and South Africa. Cuthell has run e-learning courses, and worked with MirandaNet to support the introduction of e-learning in Free State, South Africa.

Christian Dalsgaard is a PhD candidate at the Institute of Information and Media Studies, at University of Aarhus, Denmark. His research is concerned with development of learning technology on the basis of a learning theoretical approach, and he works with the design of learning materials in support of open-ended and self-governed learning activities. He has also worked with pedagogical evaluation of learning management systems and learning objects. He has published articles within the fields of learning theory, learning technology, Web-based learning, and knowledge sharing.

Caroline Daly is a lecturer at the Institute of Education, University of London. She has developed e-learning in national and international contexts, and is now a tutor and module leader on the mixed-mode master of teaching programme. She currently works in the Centre for Excellence in Work-based Learning for Education Professionals at the IOE. Her research examines the online learning experiences of practising teacher participants in asynchronous discussions. It has involved the assessment of different methodologies for gaining qualitative data that can reveal evidence of transformational outcomes brought about by participation in online learning tasks.

Olatz López Fernández is a PhD candidate from the University of Barcelona, Spain. She is writing her dissertation about the use of digital portfolios in higher education as an alternative methodology of assessment and learning. She comes from an interdisciplinary background of psychology, pedagogy and audiovisual communication (UB). Through the ILET project, she is becoming an expert on intercultural issues in educational technology, and she is doing her doctoral research with a fund from the Ministry of Education and Science of the Government of Spain in the Institute of Educational Sciences and GREAM research group (UB).

Ian W. Gibson holds the Vincent Fairfax Family Foundation Chair in Education - Teacher For The Future - at Macquarie University, Sydney Australia. He teaches leadership, technology, and research to graduate students from a curriculum/sociological perspective. His research agenda focuses upon the transformational impact of technology on learning and leading. He has received awards, honours, and grants for publications and presentations on technology use, leadership, and global learning. He presents his research through publications, international presentations, and keynote addresses, most recently in Denmark, Spain, UK, and South Africa. He is the president of the Society for Information Technology and Teacher Education (SITE). In a previous position, Ian was the chair of the Department of Educational Leadership at Wichita State University, Kansas (USA).

Dwayne Harapnuik is a lecturer for the Faculty of Education in the Department of Educational Psychology at the University of Alberta, Canada. He has been involved in instructional technology and Web-based instruction since the early 1990s. Harapnuik is also involved in researching constructivist learning environments in early childhood education through the investigation of the project approach to learning in a homeschool setting. In addition to his academic pursuits, he is involved in technology integration consulting in the private sector, and offers his technological expertise as an acting chief technology officer for an international mission and relief organization.

Simon B. Heilesen is senior lecturer in Net Media at the Department of Communication, Journalism and Computer Science, Roskilde University, Denmark. His principle research interests are hci-design and communication planning for the World Wide Web, and learning and collaboration in net environments. He has been studying and practicing net based/supported education since the early 1990s, first at Copenhagen University, later in the master's programme in computer mediated communication at RU. He is a member of various RU and national committees for the development and dissemination of information and communication technology in higher education. For more information, please visit http://www.ruc.dk/~simonhei.

Sisse Siggaard Jensen is a senior lecturer in Net Media at the Department of Communication, Journalism and Computer Science, Roskilde University, Denmark. Her principle research interests are reflective practices, social interaction, collaboration and interpersonal communication, knowledge sharing, knowledge communication, identity, avatars, and organisational games on the net. She has been studying and practicing net based/supported education for more than 15 years, both at various Danish universities and as a senior consultant at the Danish National Centre for Technology Supported Learning. For more information, please visit http://www.ruc.dk/komm/Ansatte/vip/Sisse/.

Maarten de Laat conducts research on e-learning in both educational and organisational contexts. His work covers networked learning, CSCL, ICT, communities of practice, social learning, work-related learning, and knowledge management. Besides working for the e-Learning Research Centre at the University of Southampton, UK, he works for the Centre for ICT in Education at IVLOS, University of Utrecht, The Netherlands. He is co-founder of Knowledge Works, a software company that develops software to support learning and knowledge management, and he facilitates a Dutch online workshop on the foundations of communities of practice in collaboration with Cpsquare.

Vic Lally is currently head of education and training for The Mental Health Foundation (UK). Lally's main research interests are in learning and teaching, education, research methods, and theories of learning. He is an academic with over 25 years experience in the field of education. His recent work is in the context of computer supported collaborative learning and networked learning. He also has interests in the philosophy and ethics of education, and the cultural and political contexts of learning. He is based in Scotland. Lally also works extensively with colleagues at the Beijing Normal University in China.

J. Ola Lindberg is a PhD candidate in education at Umeå University, Sweden, currently employed at the Department of Education, Mid Sweden University. His main interest lies in distance-based teacher training programmes supported by ICT. Lindberg is writing his doctoral thesis together with Anders D Olofsson, Umeå University, Sweden. In the research conducted with Olofsson, they depart from a philosophical hermeneutical approach aimed at understanding social processes of teaching and fostering. In distance-based teacher training programmes, they focus on how teacher trainees negotiate meaning in educational online learning communities. He has contributed book chapters, conference-papers, and journal articles on this specific topic.

Rema Nilakanta is a senior PhD candidate in the Department of Curriculum and Instruction at Iowa State University, USA. Her research interests lie in the area of design of online systems that support collaborative and democratic learning. She is also interested in changes, organizational and individual, that take place with implementation of collaborative technologies. Nilakanta is currently working on her dissertation, which deals with the design and development of CIT eDoc, an electronic portfolio system for PhD students in the curriculum and instructional technology program.

Anders D. Olofsson is a PhD candidate in education at Umeå University, Sweden. Olofsson is writing his doctoral thesis together with J. Ola Lindberg, Mid Sweden University, Sweden. From a philosophical hermeneutical approach, his research is aimed at understanding the meaning of social processes of teaching and fostering, establishing the meaning of, for example, moral, democracy, learning and teaching in educational online learning communities situated within distance-based teacher training supported by ICT. During his doctoral programme, he has contributed book chapters, journal articles, and papers to conference proceedings on this specific topic. Olofsson is a member of the Swedish research group Learning and ICT (LICT).

Norbert Pachler is associate dean of initial and continuing professional development and co-director of the Centre of Excellence in Work-based Learning for Education Professionals at the Institute of Education, University of London. He supervises, and has published widely, in the fields of FL education and new technologies. His research interests include (comparative approaches to) FL education and policy, initial teacher education, teachers' professional development, new technologies and educational policy. He is joint editor of the *Language Learning Journal* and of *German as a Foreign Language*.

Pirkko Raudaskoski holds a PhD in applied language studies (English, from Finland), and an MSc in knowledge based systems (artificial intelligence, from UK). She also has a docentship (educational technology: communication and interaction) in the Department of Education, the University of Oulu, Finland. Her present working site is the Department of Communication at the Aalborg University, Denmark, where she works as an associate professor. She has interdisciplinary research interests: understanding computers and practice, analysing videoed interactions, doing discourse studies. Her previous research projects include Finnish Sign Language interaction and communication between "native" and "non-native" English speakers.

Eugene S. Takle is a professor in the Department of Agronomy and Geological and Atmospheric Sciences at Iowa State University, USA. His research emphasizes the use of models of the atmosphere for studying regional climate and turbulent flow through vegetation, such as tree shelterbelts. He has also developed procedures for forecasting specific roadway weather conditions such as frost. His emphasis in instruction has been on the use of the Internet as a tool for developing interactive instructional materials.

Laura Zurita is a senior PhD candidate, currently working on her dissertation in the Department of Communication, Aalborg University, Denmark. Her research interests lie in the area of collaborative and democratic learning in intercultural settings, as she is interested in the use of information and communication technologies for development. Zurita is also a consultant in international projects about citizens' involvement and deliberative methodologies. Through the ILET project, she is becoming an expert on intercultural issues in educational technology.

Index

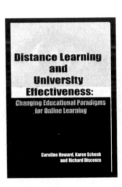